The Nature and Nurture of Love

The Nature and Nurture of Love

From Imprinting to Attachment in Cold War America

MARGA VICEDO

THE UNIVERSITY OF CHICAGO PRESS   CHICAGO AND LONDON

MARGA VICEDO is associate professor in the Institute for the History and Philosophy of Science and Technology at the University of Toronto.

The University of Chicago Press, Chicago 60637
The University of Chicago Press, Ltd., London
© 2013 by The University of Chicago
All rights reserved. Published 2013.
Printed in the United States of America

22 21 20 19 18 17 16 15 14 13   1 2 3 4 5

ISBN-13: 978-0-226-02055-6 (cloth)
ISBN-13: 978-0-226-02069-3 (e-book)

Library of Congress Cataloging-in-Publication Data

Vicedo, Marga.
    The nature and nurture of love: from imprinting to attachment in Cold War America /
Marga Vicedo.
      pages; cm
    Includes bibliographical references and index.
     ISBN 978-0-226-02055-6 (alk. paper) — ISBN 978-0-226-02069-3 (e-book) 1. Attachment
behavior in infants. 2. Mother and child. 3. Love, Maternal. 4. Attachment behavior.
5. Imprinting (Psychology) I. Title.
    BF723.A75V53 2013
    152.4'1—dc23   2012043198

TO MY PARENTS,
MARGA CASTELLÓ MARTÍNEZ AND JOSÉ VICEDO VIDAL,
WITH MUCH LOVE AND GRATITUDE

EN EL MEU COR, SEMPRE.

Contents

Introduction

David and Monica have "adopted" a robot child from the Cyber-tronics Corporation. The robot child, also named David after his human "father," is the most advanced robot of all time: not only can he think; he can feel emotions. David is capable of love. How is that capacity turned on? Monica decides to bind him to her through an "imprinting protocol," an "irreversible" process that makes the robot child's love for mother "sealed, in a sense hardwired." Like an infant imprinted on a human mother, David becomes attached to her and dreams of having her human love. The time is the indefinite future. The place is Steven Spielberg's 2001 movie *A.I.*

Mother has always mattered, but in this scene she is determinant. Mother love literally has the power to turn a robot into a child, a machine into a human, a mechanical object into a sentient being who loves and yearns for love. By imprinting herself on the child-robot, the mother awakens his ability to love. Thanks to his mother's love, he becomes human.

Though that is science fiction, the science of mother love presented in *A.I.* draws heavily on views about human nature and mother love prevalent in American society since the mid-twentieth century. In this vision, emotions are central to humanness, and mother love awakens the capacity to love that is so essential to being human and humane. The notion that children's early relationship with their mothers will determine their future personalities has become ubiquitous. In practically any story or movie, we find that the problems of the protagonists—from fear of romantic commitment to serial killing—stem from their troubled relationship with their mothers during childhood. Behind every adult vice lies a mother lacking in virtue, a mother who was unwilling or unable to devote all her love and attention to her infant. Moreover, in the wider society, this view influences personal decisions, social expectations, and public policies about custody cases, adoptions, orphanages, and child care in general.

How did we come to hold these views about the central role of emotions and the determinant power of mother love over an individual's emotional development? What social and scientific conditions made it possible for this vision of mother love to develop? What scientific research justified that vision? And what do these ideas entail for children and mothers?

In an effort to answer these questions, this book examines scientific views about children's emotional needs and mother love in the United States from World War II until the 1970s. My central claim is that during that period prominent researchers from various fields of study established the view that emotions are an integral part of the self and that mother love determines an individual's emotional development. One theory in particular played a key role in the establishment and permanence of those views: John Bowlby's ethological theory of attachment behavior. Although this was not the only theory that put forth maternal care and love as the cradle of the emotional self, it became the most enduring and successful one. I do not claim that attachment theory was the only or even the dominant paradigm of child development at this time. Indeed, I show that many scientists rejected the ethological theory of attachment behavior. But Bowlby's ethological theory of attachment became one of the most influential psychological theories of the twentieth century.

Although some psychologists have criticized certain aspects of it, in the early twenty-first century attachment research remains a major area of scientific study and social interest. Historians have noted its impact on adoption policies and on child-care advice and practices. It has sparked the development of other influential theories such as John Kennell and Marshall Klaus's theory of maternal bonding.[1] Sociobiologist Sarah Hrdy upholds Bowlby's views as a major breakthrough in evolutionary studies of motherhood. Philosopher Martha C. Nussbaum uses attachment as the foundation for her work on the contested relation between emotions and rationality. Today "attachment disorder" has entered the medical vocabulary and, although not yet recognized by the American Medical Association, has its own criteria for diagnosis and therapy.[2] Yet only a few decades ago, the theory of attachment did not exist.

The story of attachment theory began with a pressing question at the end of World War II: What do children need if they are to develop into emotionally healthy individuals and good citizens? Worried about the devastating effects of war on children, the World Health Organization commissioned a study to find out what they must have to become physically and emotionally healthy. British psychoanalyst and psychiatrist John Bowlby (1907–1990) was selected for this task. Relying on psychoanalytic studies of maternal deprivation as well as his own studies on child delinquency, Bowlby identified the mother as the psychic organizer of her child's mind. Children without mother love, he claimed, grow up to be affectionless criminals, psychopaths, or neurotic, aggressive, oversexed, and anxious individuals. Furthermore, in his view infants have an innate or instinctive need for mother love.[3]

To defend his ideas about the instinctual need for mother love, Bowlby adopted Konrad Lorenz's notion of imprinting. Imprinting is the process whereby some species of birds follow and become attached to the first object or subject they see after hatching. Synthesizing ideas from psychoanalysis and ethology, in 1958 Bowlby presented his ethological theory of attachment behavior as a new paradigm in child development. According to this theory, natural selection has provided children with a set of fixed-action patterns, like crying, sucking, and smiling, all designed to build an attachment between baby and mother.[4] Bowlby relied on psychoanalyst and social worker James Robertson's observations of children separated from their parents in hospitals. He also appealed to University of Wisconsin psychologist Harry Harlow's famous experiments involving rhesus monkeys raised with surrogate mothers made of cloth or of wire. According to Bowlby, Harlow's work on monkeys showed that maternal care in infancy is essential for adults' sexual adjustment and mental health. Later Mary Ainsworth, a psychologist from Johns Hopkins University, presented observational studies of mothers and children carried out in Uganda and Baltimore as well as experimental work that she claimed supported the ethological theory of attachment behavior. For Ainsworth and Bowlby, a child's secure or insecure attachment to his mother affects all his relationships later in life. That is, they turned this theory about the mother-child relationship into a theory of personality development.

Most historical accounts and attachment theorists to date present Bowlby and Ainsworth as the cofounders of the theory and indicate that

Harlow and Lorenz provided the animal research that proved the instinctual nature of maternal attachment.[5] Yet in exploring how scientists studied the inner needs of infants, the biological basis of human conduct, and the factors shaping a person's emotional life, I have discovered a more complex story.

Contrary to Bowlby's own assertions that attachment theory was based on the convergence of results from investigations about infants, birds, and monkeys, I argue that the theory's success was due to the cross-disciplinary alliance between Bowlby and Lorenz, who relied on each other's support to bolster their views about the biological basis of human behavior, in the face of considerable criticism. In addition, the theory's main architects, Bowlby and Ainsworth, made a concerted effort to present a united front and downplay the importance of their critics.

I present my historical analysis in three parts. Part 1, "From Imprinting to Attachment," focuses on the early context of views about the mother-child dyad during and after World War II and examines how Bowlby developed his views about attachment by adopting Lorenz's views on instincts and imprinting. It also explores the reception of this view in the context of postwar debates about gender roles and working mothers. Chapter 1, "Mother Love as the Cradle of the Emotional Self," explains how war and postwar readjustment fostered concern about emotions, which increased scientific and social interest in mothers' role in their children's emotional development. This context stimulated and drew attention to work on maternal deprivation by a number of prominent psychoanalysts, including Anna Freud, David Levy, Margaret Ribble, René Spitz, and John Bowlby, as well as to writings on maternal overprotection by Philip Wylie, Edward Strecker, and Erik Erikson. Relying on these studies of maternal separation and deprivation in his influential 1951 WHO report, Bowlby identified the mother as the psychic organizer of her child's mind.

Chapter 2, "The Study of Instincts," analyzes Lorenz's views about instincts and imprinting in the context of his efforts to develop the new science of ethology with Niko Tinbergen. In interactions with psychoanalysts like Bowlby, Lorenz claimed that mothers' behavior toward infants is instinctual. So are infants' responses toward their mothers, once they are released through imprinting. Some birds, such as ducks and geese, follow and become attached to the first moving object they see after hatching, which is usually their mother. According to Lorenz, if they imprint on the wrong object—for example, a human—they will direct the standard social and sexual responses of their species toward humans. Encouraged by psychoanalysts' interest in his work, Lorenz extrapolated his results about imprinting in birds to humans and supported Bowlby's use of his work. Lorenz thus played a key role in turning love for the mother into a biological instinct.

Chapter 3, "Bowlby's Ethological Theory of Attachment Behavior: The Nature and Nurture of Love for the Mother," discusses the reception of Bowlby's work in the United States during the 1950s. Intense social concern about emotional pathology and about the increase in mothers working in paid jobs led to a heightened scientific and public interest in the mother-child relationship. Bowlby's ideas attracted much attention. He explicitly argued that mothers should not rely on child care but should stay home with their babies. Yet several psychologists also raised pointed criticisms. To provide stronger support for his views about children's innate need for mother love, Bowlby turned to animal research and, specifically, to Lorenz's work on imprinting. In 1958, Bowlby introduced his ethological theory of attachment behavior and claimed the authority of biology in its support. This view undervalued mothers' work and mother love, as I explain later in this introduction.

Part 2, "Challenging Instincts," shows how the biologizing of human emotions and, in particular, Bowlby's views about the essential role of mother love in an infant's development received substantial criticism from various scholars. Chapter 4, "Against Phylogenetic Reductionism: the Role of Ontogeny in Behavior," presents the objections raised by comparative psychologists. Daniel Lehrman critiqued the explanatory value of the concept of instinct and the impossibility of isolating innate behavior with deprivation experiments. He further argued that the most fruitful way to understand an organism's behavior was to analyze its ontogenetic development, which reveals that the final phenotypic result is a product of the individual's genetic endowment, environmental influences, training, and contingent history. In this picture, maternal behavior is not an innate schema automatically "released" by a smiling baby, as Lorenz and Bowlby claimed. In addition, British animal researcher Robert Hinde exposed the weaknesses in Lorenz's hydraulic model of instinctual energy and the concept of drive. Finally, I examine how several animal researchers criticized Lorenz's views about the generality and fixity of imprinting. Although Lorenz tried to address those criticisms, by the late 1960s many biologists rejected his theories about the existence of innate releasing mechanisms of behavior, like the one between child and mother, as well as his views about instinctive behavior, imprinting, and the determinant role of early experiences on an animal's adult behavior.

Chapter 5, "Psychoanalysts against Biological Reductionism," examines how psychoanalysts reacted to Bowlby's attempt to unify psychoanalysis with ethology. Beginning with Freud's views about instincts, I point out some key resemblances between the psychoanalytic and ethological models. Despite these similarities, however, the aims of the two enterprises were radically different. Ethologists focused on fixed-action patterns, the observable, stereotyped behavior typical of every organism from a particular species. In contrast, as major Freudian psychoanalysts Anna Freud, Max Schur, and René Spitz pointed out, by focusing only on behavior and by reducing instinctual drives to biological instincts, Bowlby's theory had eliminated the subjective aspect of behavior and emotion that was central to a psychoanalytic account of the mind.

Chapter 6, "Primate Love: Harry Harlow's Work on Mothers and Peers," examines Harlow's famous experiments involving rhesus monkeys raised with surrogate mothers made of cloth or of wire because Bowlby used them to buttress his belief in children's need for maternal love. This research showed that even when the infant monkeys were "fed" by a bottle attached to the wire mother, they always clutched the soft cloth mother and ran to her when they were afraid. Harlow thought his experiments showed that a machine, and perhaps even a father, could substitute for the mother. But his later experiments showed that the artificial mother substitutes had produced monsters: sexually inadequate males and abusive mothers. According to Bowlby, and as reported in many subsequent historical accounts, Harlow's work thus supported Bowlby's views about the infant's need for mother love. But Harlow himself ultimately came to a very different conclusion. Following Harlow's experiments over the years, my analysis reveals a complex story about the role of mothers and peers in monkeys' lives. At least in the life of a rhesus monkey, Harlow showed that one can make do with a variety of loves. He also challenged simplistic visions of the origin of love as a matter of nature or nurture.

Part 3, "Naturalizing Nurture," shows how Bowlby used Mary Ainsworth's work on infants as empirical support for his theoretical views. It also examines how, in the face of sharp criticism, Bowlby and Lorenz continued to defend their conception of human nature by supporting each other. Chapter 7, "The Nature of Love: The Experimental Proof of the Ethological Theory of Attachment," examines the development of Ainsworth's views in the context of her relationship with Bowlby, showing her evolution from being Bowlby's defender to considering herself a coauthor of the ethological theory of attachment. Based on observational studies of mothers and children in Uganda and Baltimore, as well as observations of infants in an experimental procedure she called "the strange situation," Ainsworth argued that an infant's sense of security and attachment to the mother develops as a result of the quality of maternal care, especially a mother's sensitivity to her child's needs. She also followed Bowlby by presenting attachment as an evolutionary adaptation. But as my analysis also shows, Ainsworth's results did not adequately substantiate the biological basis of attachment.

In chapter 8, "Reinforcing Each Other and a Normative View of Nature," I explore how Bowlby and Lorenz, despite growing scientific criticism, continued to support each other. They thus reinforced the notion that their views about the biological roots of human social behavior had interdisciplinary support. And they continued to draw out the normative implications of those views. Then I examine Bowlby's introduction of the notion of the environment of evolutionary adaptedness, which provided a new twist in the naturalizing of parental roles, one that appealed to the evolutionary significance of the right environmental inputs in maintaining the behaviors and emotions optimally designed by natural selection.

After a summary of the main results and argument, in the conclusion I highlight some key factors that explain the success of attachment theory in the face of substantial criticism. In addition, I reflect on what my historical analysis says about its value. Since attachment theorists have often used historical reconstruction to legitimize their science, a reinterpretation of that history must bear on the status of the theory. I suggest that over the course of a few decades, attachment theorists failed to address some of the basic challenges their critics raised. I focus on the need to clarify the relations among behaviors, emotions, and instincts and the questionable uses of animal models in human psychology.

*   *   *

The history of attachment theory that I present had implications for larger developments that I can only briefly indicate here: a turn toward a biological conception of human nature and of maternal sentiments, in particular, which led to a devaluation of mother love and provided a new justification for gendered parental roles.

First, by providing an evolutionary framework for emotions, after

World War II attachment theory contributed to a renewed interest in instincts as an explanatory category of human behavior and emotions. A number of historians have argued that instincts had become devalued in different areas. In his close examination of controversies about instincts before the war, Hamilton Cravens concluded that by the 1930s students of human behavior had agreed on a framework that recognized the interaction between nature and nurture. Influenced by the nascent experimental ethos, they moved away from a concept of instinct that was both simplistic and too ambiguous to use in experimental research. Carl Degler has claimed that the interwar period saw the triumph of cultural over biological accounts of human behavior, mainly for ideological reasons. In the case of maternal instincts, Nancy Cott has argued that the dismissal of instincts reflected a wider rejection of appeals to animalistic tendencies in maternal care. She also found that social scientists asserted that mothers should follow the teachings of science. Rima Apple has explored the movement of "scientific motherhood" in detail. According to all these historians, by World War II instincts were no longer a central part of psychology and scientific visions of motherhood.[6]

But instincts remained central in psychoanalytic accounts of the human mind, and such accounts flourished during and after the war, for several reasons. The animal researchers Konrad Lorenz and Niko Tinbergen became highly successful in promoting the new science of ethology as the biological study of instincts. Psychoanalysis and especially child psychoanalysis also focused on instinctual drives. Attachment theory built on and contributed to evolutionary accounts of human behavior and, more specifically, to explanations of human social behavior and emotions that gave a large role to instincts. In doing so, this theory played a key role in a renewed biologizing of human nature, with important implications for our understanding of maternal love and gendered parental roles.

By naturalizing maternal love, I claim, Bowlby's theory of attachment behavior devalued maternal sentiments and provided a new justification for gendered parental roles. To be sure, many of Bowlby's ideas about the importance of mother love were hardly new. Views about the early significance of child experiences, the mother as the nurturer of her child's emotional disposition, and the social importance of raising children with the right character were all elements in American political and family history since the end of the eighteenth century, when economic and social changes increased the separation of men in the public realm and made women keepers of the hearth. As historians such as Ruth Bloch and Jan Lewis have shown, the power of mother love in shaping a child's mind is "the kernel of the emotionology of motherhood" in the United States.[7] So the belief in the power of mother love did not first arise in the middle of the twentieth century. Furthermore, the idea that children have emotional needs that only parents, and more specifically mothers, can meet was not a postwar invention. In her study of Boston's Judge Baker clinic for problem children, Kathleen W. Jones convincingly argued that "child guiders cast the troublesome child as a bundle of emotional needs that could be met only through the establishment of satisfying relationships with parents."[8] In addition, the appeal to biology and evolution to justify different roles in society for men and women and, more specifically, different parental roles dates back at least to the mid-nineteenth century, as historians like Charles Rosenberg, Carroll Smith-Rosenberg, and Cynthia Russett have shown.[9]

Given the deep roots of these ideas, it might be tempting to see the story of attachment theory as simply a continuation of views already well established in American society and science before World War II. But we must resist that temptation. Even if the ideas that attachment theorists presented were the same as previous ones, it would still be important to understand how they continued to have power in different socioeconomic and scientific contexts. Inertia does not explain why beliefs endure. Maintaining the status quo in any area requires constant negotiation and justification in the face of changing social circumstances. When the sociocultural environment changes, beliefs may no longer survive.

More important, I argue here that attachment was not a mere continuation of old ideas about mother love under a new name. Bowlby's biologizing of children's emotions and his postulating their instinctual need for maternal love were new ideas in child development. In his early writings, Bowlby, like many other authors, argued that children need maternal care and love to develop into healthy individuals. But later he presented children's emotional needs as biological instincts. Furthermore, he then introduced a new concept, the environment of evolutionary adaptedness, to argue that natural selection had constructed a mother-child dyad in which the instinctual needs of each half work in tandem and whose disruption has dire consequences.[10] In Bowlby's view, children are organisms constructed by natural selection to function properly only with the right amount and type of maternal love. In addition, by enlisting Lorenz's support, he laid claim to the authority of biological science to back up his views about human nature and emotions.

Bowlby's views about children's instinctual need for mother love exerted a strong emotional demand on mothers and contributed to an increasing discourse of mother blame. Attachment theorists claimed maternal love plays a key role in a child's development. Yet by turning mother love into an evolutionarily programmed behavior and emotion, proponents of attachment theory left maternal sentiments outside the realm of moral value and praise.

My interpretation of the devaluation of mother love agrees with the analysis provided by Rebecca Jo Plant in her recent book *Mom: The Transformation of Motherhood in Modern America*.[11] Plant has shown that transforming mother love into a psychobiological drive was an important last step in dismissing a conception of moral motherhood that had fueled maternalist discourses at the turn of the twentieth century. Such discourses, which appealed to maternal virtues to ground women's entitlements in society, had recognized that motherhood was a hard task. But as Plant notes, by the postwar period, mother love was increasingly posited as a natural and effortless feeling. Plant does not examine research on children or attachment theory, but my conclusions about the importance of these scientific developments dovetail nicely with her views about the devaluation of motherhood after World War II. Bowlby's attachment theory continued the rejection of moral motherhood and, furthermore, introduced a new justification for gendered parental roles.

Although some psychological and psychoanalytic research that contributed to the discourse of mother love and mother blame in the postwar era did not survive into later periods, Bowlby's ethological theory of attachment behavior actually became more widely accepted in psychology and the wider culture. His appeal to biology became crucial here, especially his notion of the environment of evolutionary adaptedness, the one in which the basic instinctual responses characteristic of the mother-infant dyad were established. The ethological theory of attachment postulates that babies and mothers are born with a repertoire of responses that evolved through natural selection in the human "environment of evolutionary adaptedness," the ancestral environment in which human beings evolved and which they are adapted to. In Bowlby's view, because the mother is the "environment" of the child, her responses must not deviate from the range of behaviors designed by natural selection and needed for the survival of her children and therefore the species. Deviations from the patterns established by natural selection will lead to "maladaptive patterns of social behavior."[12] According to Bowlby's formulation, nature and nurture had reached a functional balance whose disruption could only have detrimental effects. This position provided a functionalist justification of gender roles with strong deterministic consequences. The natural became the normal because of its adaptive value. It was also the good because it worked to ensure society's survival. This vision of a mother-infant dyad designed by evolution entails a normative model of motherhood. In this picture, our past evolutionary history has set clear constraints on our capacity to develop new ways of parenting without disrupting the optimal natural order.

Yet this particular way of constructing human nature and viewing human emotions was based on a specific view of evolution that received pointed criticism and was fostered by particular sociocultural circumstances. Recovering the history of those criticisms is important for assessing the validity of scientific ideas that inform our beliefs and shape our feelings.

Mother love has been nurtured in diverse ways throughout the ages, as have other human emotions. Recently scholars have devoted extensive attention to the historical significance of the attitudes and standards that different societies maintain toward emotions. By placing emotions in their sociohistorical context, their work reveals how societies shape people's feeling and how those feelings influence a variety of areas. Specific societies encourage and discourage emotions and affects in different ways. The analysis in this book aims to show that science has played a key role in shaping our views about two basic emotions: mother love and love for the mother.[13]

Our views about those emotions are central to our understanding of the role of feelings in human conduct, our ideas about the biological underpinnings of human behavior, and our debates about parental and gender roles. Mother love and love for the mother may be natural, but societies have nurtured different views about their origin and significance, and science has played an increasingly larger role in shaping them. Revealing the contingent character of this history opens up new possibilities for thinking about our emotional selves.

Note on Terminology

Instinctive, instinctual, and innate: In this book I use these terms interchangeably, unless I am citing or using the terms that some authors employed.

Imprinted on: Sometimes one can read that a bird imprinted on Lorenz, for example, and also that Lorenz imprinted on a bird. In the literature one can find both uses and they refer to the same thing.

Ethological theory of attachment behavior: I use this to refer to Bowlby's views after 1958, when he presented a synthesis of ethology and psychoanalysis to explain attachment. But the developers of attachment theory and the literature refer to it in different ways: Bowlby's evolutionary ethological approach, for example, or simply attachment theory.

PART I
From Imprinting to Attachment

CHAPTER ONE

Mother Love as the Cradle of the Emotional Self

"How can we rear an emotionally healthy generation?" —Midcentury White House Conference

Introduction

World War II split the century, the atom, and the family and, perhaps worst of all, destroyed the faith that a war could put an end to all wars. In search of an explanation for human destructiveness, many social scientists turned to the nonrational causes of human behavior—the emotions. As they saw it, ensuring a peaceful world order required controlling human emotions, and to do this one first needed to understand them. In seeking out the source of an individual's emotional nature, scientists soon found the mother. Their reasoning went like this: Personality is a direct result of the individual's emotions; emotions are created in childhood; and children are raised by their mothers. The mother thus became the foundation of a healthy personality. But according to many experts and to others with little expertise, American mothers were creating a problem of national proportions. They were sometimes accused of smothering their children and at other times of depriving them of love. In either case, mothers turned their children into emotional cripples and therefore put the social order in danger.

How did mothers acquire so much power? What scientific research justified this vision of mother love as determinant of individual personalities and national character? To answer these questions, this chapter explores the development during the 1940s of scientific views about children's emotional needs and about the role of mother love in shaping their personality. Concern for emotional and mental health preceded the war, as did mother blame. Kathleen W. Jones's study of the interwar period demonstrates that mother was a convenient scapegoat for children's behavioral and emotional problems in that era as well. From psychologists like John Watson to child guidance workers, children's mental problems were traced back to their alleged doting or emotionally negligent mothers.[1] However, I suggest here that the war also played a key role by fostering widespread concern about the role of emotions in human behavior. In turn, the interest in the formation of the emotional self, I claim, was crucial to the success of child psychoanalysis around and after World War II. Both factors encouraged research on mother love and love for mother.

American historians have shown how authors like popular writer Philip Wylie and psychiatrist Edward Strecker blamed "moms" for their sons' immaturity. For Mary Jo Buhle, "momism" helped instigate a misogynist discourse in the postwar era. Rebecca Jo Plant sees momism as part of a larger move to reject an older vision of moral motherhood that had fueled maternalist discourses at the turn of the century. Plant also pays attention to some of the literature on maternal deprivation and notes the shift toward a conception of maternal love as a psychobiological drive.[2]

Here I focus on the scientific research on child development that supported those visions of maternal care and love. I examine how discourses that blamed mothers for loving children too much and loving them too little overlapped during many years, but toward the end of the 1940s I identify stronger concern about the absence of mother love in an infant's early years. In this area, I analyze the views of prominent researchers on maternal deprivation, such as Anna Freud, Margaret Ribble, René Spitz, David Levy, Therese Benedek, Erik Erikson, and John Bowlby. These authors presented mother love not merely as something children desire or benefit from but as an instinctual need. Yet despite their constant appeal to infants' innate emotional needs, most of the authors discussed here did not clarify what they meant by instincts. By the end of the decade, in the popular version of a highly influential report for the World Health Organization, John Bowlby appealed to the work of animal researchers to support the idea that maternal deprivation has catastrophic consequences because the mother is the child's psychic organizer.

Becoming Emotional

Consider these two sets of data: During World War I, thousands of men were rejected for military duty because of their poor performance on intelligence tests; during World War II, almost three million men were rejected for being emotionally unstable.[3] In the two periods, these results became central points of reference in discussions about what was wrong with American society. In contrast to the focus on intelligence and intellectual capacities prevalent around World War I, around World War II the alarming results about the troubled emotional character of Americans helped consolidate the perception that the country needed to care about its citizens' emotions. This concern was not wholly new, but during the war and its aftermath one can discern an increasing emphasis on emotional factors in shaping human behavior and causing social problems. Although it is beyond the scope of this book to unravel all the complex factors underlying this shift, let me point out several factors that contributed to it.

The war helped spotlight emotions by making it clear that intellectual improvement was not enough to ensure peaceful social relations and, on its own, might not even be desirable. The rise of Nazism in one of the most highly educated societies, and the horrifying systematic, mechanized, and calculated murder of millions, had made that frighteningly clear. After the war, the very signs of scientific mastery—like the atomic bomb—fueled fears of a society that was becoming more and more technologically sophisticated, but also dehumanized.[4]

In this context, many social commentators were disturbed by the perception that societies were improving their intellectual and technical capacities at the expense of their emotional and moral development. After a war that had depended heavily on machines and technical achievements, building a peaceful democratic order seemed to require balancing the advancement of society in other realms. Edward A. Strecker, in his 1944 presidential address to the American Psychiatric Association, noted that it was "unlikely that democracy will fulfill its destiny by grace of technocracy alone, even though scientific technical achievement has been truly amazing."[5] As Harvard psychologist Gordon Allport saw it in 1947, "the outstanding malady of our time" had many expressions, but "the underlying ailment" was "the fact that man's moral sense is not able to assimilate his technology." To prevent the triumph of a dehumanized, mechanical world, Allport called for more attention to "problems of human affection and the conditions for its development."[6]

The war also contributed to ongoing interest in understanding human nature and, specifically, clarifying whether aggression is an ineradicable part of human behavior. "War is not inevitable and not part of human nature": so said the manifesto signed by a dozen psychologists at the celebration of Armistice Day in 1937. They represented the consensus of their fellow psychologists, who had passed their verdict in democratic fashion. The American Psychological Association asked its members: "Do you as a psychologist hold that there are present in human nature ineradicable, instinctive factors that make war between nations inevitable?" Of the 528 members, 378 cast a vote, which stood as follows: no, 346; yes, 10; unclassified, 22. Most psychologists, including hard hereditarians such as H. H. Goddard and S. J. Holmes, thought war is not instinctive or inevitable. But many of them thought there are instinctive factors in human nature that often get derailed in the madness of war. World War II once again brought the question to the fore: "Is War Instinctive—And Inevitable?" asked the biologist Julian Huxley in the *New York Times*. Others continued a discussion that would only become more heated during the tense years of the Cold War.[7]

This concern about human instincts also rekindled interest in the emotions. In most scientific and popular accounts up to World War II, emotions and instincts were used almost interchangeably. In his foundational text of psychology, pragmatist psychologist William James noted: "Instinctive reactions and emotional expressions shade imperceptibly into each other. Every object that excites an instinct excites an emotion as well." Many psychologists of different orientations agreed. In the words of John Watson, the father of behaviorism: "There is no sharp line of separation between emotion and instinct." Watson's nemesis, Harvard psychologist William McDougall, also correlated emotions with instincts, identifying several primary pairs of instincts-emotions.[8] Thus the desire to understand the instinctual roots of aggression also encouraged debates about emotions.

The war and its aftermath heightened worries about the emotional life of children in particular. In England, children were taken from their families to protect them from the bombs. Although these children had all their basic physical needs met in the wartime nurseries, many suffered terrible emotional trauma. The news about children who had lost their parents in concentration camps was even bleaker. "Will the walls be here tomorrow?" asked one child refugee in an English nursery. In Europe, thousands of children were left homeless and surely heartless by the war and its aftermath. Numerous studies, including the pioneering ones by psycho-analyst Anna Freud, examined the emotional damage the war inflicted on children. Although American children were not directly affected, people in the United States were also concerned: "What has war done to the children of the world, and what can we do about it?" Discussion groups in the United States would tackle the question, announced Dr. Fremont-Smith, director of the Josiah Macy Jr. Foundation, at the 1947 World Mental Health Congress. One year later, the United Nations General Assembly Universal Declaration of Human Rights said that "motherhood and childhood" deserved "special care and assistance."[9]

At home, postwar difficulties in readjusting to work and family life underscored the importance of emotions. American historians have documented the tremendous repercussions postwar readjustment had on Americans' emotional life. After confronting the brutalities of war, the GIs came home and needed to adjust to factories, corporations, and families that had undergone deep changes in their absence.[10] As Rebecca Jo Plant has noted, the literature on helping veterans and their families "portrayed readjustment in psychological terms, as an emotionally fraught process that could result in serious difficulties if mishandled."[11] Regarding family life, historian Nancy Cott has shown how the literature on readjustment "stressed that veterans hoped and deserved to recapture the traditional marital constellation, with the father/husband the provider and protector, and the wife/mother the sympathizer and nurturer." However, the domestic bliss that many men had dreamed of in battle could turn into a nightmare as they returned to families that were not quite the same as when they left.[12]

Women's lives had undergone dramatic changes as well, and for them readjustment also proved difficult. During the war, women entered the workforce in unprecedented numbers. Afterward many families moved to the suburbs, also in unprecedented numbers. Women were expected to leave their jobs and focus on being supportive wives and nurturing mothers.[13] At least for educated middle-class women, these developments exacerbated the conflicting feelings and anxieties they were already experiencing during the war years. As sociologist Mirra Komarovsky showed in a 1946 study of college women, seniors commonly faced contradictory expectations about their adult roles: homemaker and "career girl." Confronted with these competing ideals, they suffered "from the uncertainty and insecurity that are the personal manifestations of cultural conflict." Because these roles required different personality traits, the conflict left young women in "bewilderment and confusion." Over the next decade, numerous studies by sociologists as well as women's narratives documented their emotional turmoil. A 1947 issue of *Life* magazine identified the problem as "The American Woman's Dilemma." According to the editors, a growing number of women were confused and frustrated. They felt conflicted about the new emphasis on women's duties in the home and the reality of their involvement in public activities.[14]

Even fiction showcased the emotional problems of readjustment. William Wyler's 1947 Oscar-winning movie, *The Best Years of Our Lives*, explored the problems of going from soldier to citizen in a country with new economic and social expectations. One of the returning veterans comes back to an estranged wife and can only get a job selling cosmetics. Another one returns to a high position in a bank, but he is forced to deny loans to veterans who have no collateral. And a handless veteran feels frustrated because he is dependent on his family. All three men suffer emotional problems owing to the strains of readjustment.

Besides coping with new aspects of family life, men and women had to adjust to new expectations in the workplace. Here too, social commentary centered on the significance of personality traits and emotional maturity. In 1949, the *Report of the Study for the Ford Foundation on Policy and Program* presented an alarming picture:

> No census can show how many persons in our society labor under the disabling effect of inadequate emotional adjustment. The estimates vary widely; some authorities regard emotional maladjustment as the most characteristic and widespread ill of our civilization. In a small percentage of instances this maladjustment takes the form of violent social disorders such as crime, delinquency, and insanity. In the great majority of cases it is revealed in illness, in unstable family life, in erratic and unproductive work habits, and in inability to participate effectively in community life. This maladjustment makes people unable to have satisfactory relations with their fellows, unwilling to cooperate adequately, and unable to compete successfully.[15]

Thus emotional maladjustment seemed to threaten the survival of the patriarchal family, the cohesiveness of the social fabric, and the success of American capitalism. Emotional maturity probably had long been considered important for satisfactory personal relations, but many commentators now also emphasized its significance for social and political stability. This trend fits well with the tendency to psychologize social and cultural reality that Richard Pells identified as a characteristic of the 1940s and 1950s and the postwar "romance with psychology" examined by Ellen Herman.

During this period, emotional and psychological problems such as anxiety, insecurity, immaturity, and imbalance were taken to contribute to major social ills. The spectacular growth of the fields of psychiatry, psychology, and psychoanalysis attests to the increasing perceived need to attend to the mental health and emotional problems of the American people. In response to such grave problems, the federal government took action. It made the citizen's mental health a major priority by passing the 1946 National Mental Health Act, then created the National Institute of Mental Health (NIMH) in 1949. In 1949 the NIMH budget was $9 million, in 1959, $50 million, and in 1964, $189 million.[16]

Even the survival of democracy came to be seen as an emotional problem. "The real difficulties with democracy . . . are emotional," Franz Alexander, founder of the Chicago Psychoanalytic Institute, stated in 1942.[17] Contemplating a suicidal third world war enabled by the advent of the atomic bomb, British psychiatrist and psychoanalyst John Bowlby concluded his reflections on psychology and democracy with these words: "The hope for the future lies in a far more profound understanding of the nature of the emotional forces involved and the development of scientific social techniques for modifying them."[18]

His fellow Americans agreed that the construction of a good society and a sound world order depended not only on economic resources, technological progress, and military might but also on emotional maturity. When experts from different fields and tendencies met in 1950 at the Midcentury White House Conference to address the key problems of the day, they raised one central question: "How can we rear an emotionally healthy generation?"[19] Well, who rears every generation?

Between Overprotection and Deprivation:
The Mother-Child Dyad Takes Center Stage

Both in the scientific literature and in popular culture, there has been a long-standing tendency to see emotions as the realm of women and, specifically, to view mothers as the emotional providers in child rearing and mother love as the source of the social sentiments. Nineteenth-century evolutionary thinkers like Charles Darwin and Herbert Spencer said mother love is the original and primary source of altruism in the natural and social worlds. Prominent early twentieth-century psychologists William James, Stanley G. Hall, and William McDougall selected mother love as the cradle of all good sentiments. Some anthropologists followed suit. For example, in his massive work *The Mothers: A Study of the Origins of Sentiments and Institutions,* Robert Briffault proposed that all social feelings derived from the maternal instinct.[20]

In postwar America, the search for the genesis of the emotional self focused attention on the field that placed the most emphasis on the determinant character of the early years and the role of emotions during childhood: psychoanalysis. As historians have documented, psychoanalysis experienced a period of revival during this time for a variety of reasons, including the desire to understand the irrational aspects of human behavior and the immigration of many European Jewish psychoanalysts to the United States and England.[21]

The concern about the formation of emotions as well as this renewed interest in psychoanalysis, and specifically in mothers' importance in emotional development, played a major role in the rise of child analysis. An indication of the expansion of this field was the founding in 1945 by Anna Freud, Heinz Hartmann, and Ernst Kris of *Psychoanalytic Study of the Child*, an annual journal that enjoyed immediate success. As René Spitz wrote to Anna Freud on August 4, 1947, "The interest in the psychoanalytic approach to the psychology of infancy is growing enormously. Lectures like those of which I am sending you reprints attract huge crowds of psychiatrists, pediatricians, obstetricians, nurses, and social workers. Lay audiences also participate with interest and very intelligent questions."[22] In turn, the rise of child analysis pushed mothers to center stage. In the United States, interest in the formation of the emotional self, psychoanalysis, and the mother's role in a child's emotional life were deeply intertwined.

Psychoanalysis posits the mother as the source of one's ability to love. According to the field's founding father, Sigmund Freud, love is born at the mother's breast: "Love has its origin in attachment to the satisfied need for nourishment." In this relationship, Freud claimed, "lies the root of a mother's importance, unique, without parallel, established unalterably for a whole lifetime as the first and strongest love-object and as the prototype of all later love relations—for both sexes."[23] Taking this first relationship as a template, the child learns to generalize this love to other persons. Many of Freud's followers, especially analysts from the Budapest

FIGURE 1.1. Children with gas masks in the Hampstead Nurseries run by Anna Freud and Dorothy Burlingham, London, 1940. Freud Museum, London.

school of Sándor Ferenczi and Freud's daughter Anna Freud, further emphasized the significance of mother love for the child's emotional well-being. Anna Freud's work with children during the war helped support these views.

After training in psychoanalysis with her father, Anna Freud (1895–1982) started a school with her friend Dorothy Burlingham and lectured on psychoanalysis at various schools in Vienna. In 1936 she published *The Ego and the Mechanisms of Defense*, a major text in what became known as ego psychoanalysis. This approach focused on the developmental stages of the ego's formation and on the adaptive and maladaptive effects of the ego's defenses (e.g., repression, regression, projection, and sublimation).[24] Anna Freud and her family moved to England in 1938 to escape the Nazis, and in 1941 she and Burlingham opened the Hampstead Nurseries outside of London to take care of infants and children whose parents could not do so during the war. The nursery had about a hundred children. When it closed, Anna Freud opened the Hampstead Clinic in 1952.

Anna Freud and Burlingham observed that the mothers' behavior in situations of danger greatly influenced their children's reactions. If a mother was calm, her child remained calm. If she was distressed, her child panicked. They also discovered that children who were abruptly separated from their mothers during the war suffered damaging emotional consequences. Even when nursery children were well provided for in material and physical terms, they often regressed to earlier stages of development, acquiring new fears and insecurities. During World War I, Anna Freud pointed out, people had realized that things like fresh fruit are not luxuries; children need vitamins for adequate physical well-being. In a similar way, she noted that World War II drove home the point that the mother is as necessary for psychic development as vitamins are for physical development.[25]

The mother provides affection as well as care, food, warmth, and cleanliness. This relationship, according to Anna Freud, is one-sided. The mother gives and the child receives. Another person can replace the mother in early infancy. But based on this primitive "stomach love," the child develops an attachment to the mother before the end of the first year. Later this attachment expands to include the father and other family members. In this way, the first relationship with the mother allows the child to develop the "ability to love."[26]

Anna Freud's views became very influential in postwar America. Her practical bent, her focus on the ordinary but often sad experiences of English children affected by the war, her clear writing and modest manner, and her stature as her father's daughter made her immensely popular.[27] An indication of American interest in her work is that the Hampstead Nurseries were maintained financially by the Foster Parents Plans for War Children, a charitable organization in New York. After the war nurseries closed, but her work received support from several other American organizations. Further evidence comes from the numerous invitations she received to lecture in the United States, which she visited several times, and from news sent to her from child analysts working there. When Anna Freud lectured during a major American tour in 1949, her findings from the war nursery captured the imaginations of psychoanalysts and the public.

She found fertile ground also because some American researchers had already been studying the relationship between children and mothers. The most influential were David Levy, Margaret Ribble, René Spitz, and Therese Benedek, medical doctors and psychoanalysts who carried out clinical work and research in institutions or hospitals. Their work was practical. They used little theory or psychoanalytic jargon. And they claimed the objectivity of scientific observations and experimentation. In

FIGURE 1.2. Anna Freud (*far right*) and Dorothy Burlingham (*far left*) with coworkers and a group of children in the Hampstead Clinic, 1966. Freud Museum, London.

the postwar context of rising experimental science, those characteristics mattered greatly, and the impact of their studies transcended the psychoanalytic community.

Born in Pennsylvania, David Mordecai Levy (1892–1977) obtained an MD at the University of Chicago in 1918 and did postgraduate training in neurology and psychoanalysis at the University of Zurich. In 1927 he moved to New York and married Adele Rosenwald, daughter of philanthropist Julius Rosenwald and an active figure in child policy. Levy became chief of staff of the New York Institute for Child Guidance, professor of psychiatry at Columbia University, and a successful private practitioner. During World War II he served as consulting psychiatrist for the Office of Strategic Services. He was a consultant to many organizations, including the US Public Health Department. He was elected president of the American Orthopsychiatric Association and the American Psychoanalytic Association and a fellow of the American Psychiatric Association. A member of New York's upper class as well as a successful author of five books and numerous articles, Levy was well respected in psychiatric and psychoanalytic circles and well known outside them.[28]

One of the first psychiatrists to analyze children's emotions, Levy argued that the child has a "primary affect hunger," or "an emotional hunger for maternal love and those other feelings of protection and care implied in the mother-child relationship." The symptoms of "affect hunger," he claimed, are manifested in children "who receive maternal care and direction to a high degree in a physical and intellectual sense, though without any evidence of affection." As early as 1937, Levy used the analogy of mother love and vitamins. According to him, as with vitamins, lack of maternal love does not result in an immediate and visible degeneration of the body. Nevertheless, the long-term consequences could be equally tragic. As an example, Levy presented the case of an eight-year-old girl who had been adopted a year and a half before referral. She was an illegitimate child who had been passed from one relative to another and finally was sent to a foster home. She lied and stole. Although she appeared "affectionate on the surface," her foster parents complained that she did not seem capable of showing any real affection. As they put it, she "would kiss you but it would mean nothing." According to Levy, this is how lack of mother love disfigures children. They grow up apparently normal in health and physical appearance, but they are emotional cripples.[29]

A few years later, in a paper titled "Disorganizing Factors of Infant Personality," New York psychoanalyst Margaret Ribble emphasized the child's need "for a long and uninterrupted period of consistent and skillful psychological mothering." This need she called "'stimulus hunger' because of its peculiar instinctual quality and because of its close analogy to food hunger." According to Ribble, a "biological symbiosis" between mother and child is essential for the emotional well-being of both. This "attachment between infant and mother," she claimed, is required for the infant's "nervous integration."[30] Her paper was short on evidence, and it did not have a single reference to other literature or many specific details about her own studies. Ribble elaborated her views in her 1943 book *The Rights of Infants*. Again she presented few systematic data or details about her work but noted that she relied on observational studies she had carried out over the years in various settings. These included observations at the Boston Children's Hospital on children suffering from birth injuries and congenital disorders of the nervous system, which she contrasted with studies at the Boston Psychopathic Hospital on children with behavior disorders who did not have organic nervous diseases; two years of study at the Wagner Jauregg Clinic in Vienna, focusing on mentally ill adult patients; her study of psychoanalysis with Anna Freud during this period; and later her observations of newborn babies in three maternity hospitals in New York City. She reported having observed about six hundred healthy infants to follow their response to routine care, a group of infants with injuries, some infants in the home, and twenty premature infants. To study the child-mother relationship, she monitored twenty children from birth. In addition, she observed a group of one hundred expectant mothers.[31]

Based on these varied observations, Ribble concluded that infants have an "innate need for love, which is a necessary stimulus for psychological development." This innate need is the root of other essential developmental processes as well: "It is the first relationship of life which activates the feelings of the baby and primes his dormant nervous system into full functional activity, giving to each individual personality its original slant." Upgraded in Ribble's work from vitamin to essential nutrient, mother love was now considered vital: "Social impulses are part of our primary equipment; emotional hunger is an urge as definite and compelling as the need for food."[32]

But what exactly does the child need from mother? Good mothering, Ribble claimed, included physical care, contact with the newborn, plus "the whole gamut of small acts by means of which an emotionally healthy mother consistently shows her love for her child, thus instinctively stimulating his psychic development." The mother should provide "tender feeling—fondling, caressing, rocking, and singing or speaking to the baby," activities with "a deep significance." Ribble offered mothers "a word of reassurance." They need not worry, since the emotionally healthy mother is a good mother naturally: "These are really very simple and primitive matters in practice, even if a bit complicated in their analysis. The woman who is herself emotionally healthy soon learns, both by instinct and by observation, to know her own baby." Laying constant emphasis on the natural character of the mother-infant relationship, Ribble asserted that the emotionally sound mother gives her child the love he needs "just as naturally as she secretes milk." But the modern woman, who was sometimes not as well attuned to her infant's instincts, needed "reassurance that the handling and fondling which she gives are by no means casual expressions of sentiment but are biologically necessary for the healthy mental development of the baby."[33]

In the mid-1940s, these psychoanalytic ideas spread widely thanks to the book that became the child-rearing manual of the postwar generation: Benjamin Spock's *Baby and Child Care*. Spock was greatly influenced by psychoanalysis and twice underwent analysis himself. His circle of friend included psychoanalysts and other influential scholars close to psychoanalysis, including Margaret Mead, Erik Erikson, Kurt Lewin, Lawrence K. Frank, and Caroline Zachary. David Levy was an instructor at the New York Psychoanalytic Institute where Spock was a student in the mid-1930s.[34] Adopting the nutrition metaphor, Spock presented his own recipe for a child's adequate emotional growth: "Every baby needs to be smiled at, talked to, played with, fondled—gently and lovingly—just as much as she needs vitamins and calories."[35]

Right after Spock published his book, the view that mother love is essential food for a child's brain acquired tremendous visibility thanks to the work of émigré psychoanalyst René Spitz (1887–1974), who claimed a mother provides the emotional sustenance that is the basis of all other aspects of human growth. Born in Vienna, Spitz graduated in medicine at the University of Budapest in 1910. He credited Ferenczi with encouraging him to pursue psychoanalysis, and he sought a didactic analysis with Freud during 1910–11. He taught in the Berlin and Paris Psychoanalytic Institutes until he relocated to the United States in 1938. He trained and supervised students at the New York Psychoanalytic Institute, and in 1956 he became a visiting professor of psychiatry at the University of Colorado School of Medicine.[36]

From the mid- to the late 1940s, Spitz published several very influential papers about work he had carried out himself and in partnership with the Austrian child psychologist Katherine Wolf, who emigrated and worked at Yale University. Spitz had observed children in the nursery of a women's prison outside New York City and in a foundling home in Central America whose specific locations he never disclosed. In the prison nursery, the delinquent women took care of their children. In the foundling home, nurses fulfilled the children's material and physical needs. Spitz reported that the babies in the penal institution developed normally, but the infants in the foundling home literally withered away. They lost weight, slept poorly, and became withdrawn. Many children died and others showed severe developmental problems. Spitz portrayed these problems in a heart-wrenching 1947 film titled *Grief: A Peril in Infancy*.[37]

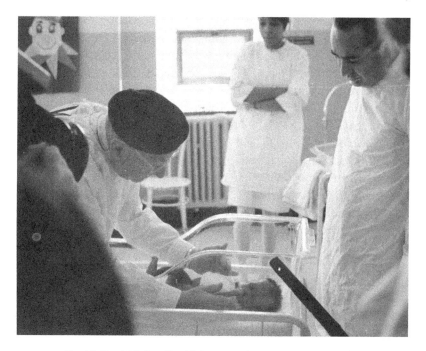

FIGURE 1.3. René Spitz with infant. René Spitz papers, box V41, folder 3, Archives of the History of American Psychology, Center for the History of Psychology, University of Akron.

Spitz used the terms "hospitalism" and "anaclitic depression" to refer to the debilitating conditions affecting children deprived of maternal care and love.[38] In an 1897 editorial in the *Archives of Pediatrics*, Floyd M. Crandall had introduced the term hospitalism to name the condition responsible for the death of many infants in hospitals. His analysis of several factors influencing the poor health of hospitalized children noted the lack of efficient care, among other factors, but he concluded that the problems were largely due to a "single cause—overcrowding." He called for more efficient nursing as well as clean air and exercise for infants in hospitals.[39] Freud used the term anaclisis ("leaning on") to refer to the "anaclitic object-choice," meaning the person infants choose for their first libidinal attachment. In Spitz's work and subsequent appropriations, both "hospitalism" and "anaclitic depression" commonly refer to the "wasting away" of an infant deprived of maternal care and love during the early months.

Spitz's views were widely discussed. One of the first students of child behavior to use film in his research, he was sought after as a speaker. His films and writings became well known and influential. "The emotional development of a baby is the 'trailbreaker' for every other development bodily or mental, Dr. René Spitz, world-famed psychiatrist, declared today," is the way the *New York Times* reported his findings in 1949. These conclusions, according to the article, were based on "twelve years of research with 1,000 babies" who were observed in institutions and homes in three countries (the United States, a Latin American country, and a European country) with the aid of comprehensive tests, including developmental tests, Rorschach tests given to the mothers, direct observations of babies, interviews with personnel in baby care, and especially films. According to Spitz, depriving a baby of the emotional satisfaction the mother provides is "tantamount to depriving an adult of every possibility of adequate contact with his environment." The result is arrested or retarded development.[40]

Less widely discussed in the general press during this period, but also respected and influential, was the work of Therese Benedek. Born in 1892 in Hungary, Benedek obtained a medical diploma in 1917 at the University of Budapest, where she attended Ferenczi's lectures. She worked as a physician and psychoanalyst in Leipzig and became a member of the Berlin Psychoanalytic Institute before moving to the Chicago Institute of Psychoanalysis in 1936. Benedek believed that motherliness helps the infant develop a "sense of confidence." That confidence preserves the mother-infant unity, and it ensures that the infant grows free from anxiety. Thus she saw a "sense of security in the relationship with the mother" as necessary for the healthy development of the ego, which could then deal effectively with other relationships.[41]

But in making mothers the cradle of all loves, these and other psychoanalysts also turned mothers into the cradle of all ills. The logic was simple: if mother love is the source of a healthy personality, disturbances in maternal love must be the source of a pathological personality as well. In their influential 1942 book *Love against Hate*, American psychiatrist Karl Menninger and his wife, Jeanetta Lyle Menninger, made that position explicit. They started with a fact: women raise most children. Thus, mothers must be the cause of their children's frustrations, gratifications, and training. They concluded: "It is, therefore, a presumptive conclusion that the patterns of emotional behavior, both of loving and of hating, are to a far larger extent than any of us realize determined not by 'the parents' but by *the mother*." So it was imperative to understand the "solemn and frightening power that mothers exert," as the Menningers put it.[42]

By situating the mother at the center of ego formation and disintegration, and by presenting mother love as the necessary element that allows an infant to develop the ability to love, child analysts contributed
to a widespread and apparently contradictory discourse that blamed
American women for all the ailments considered pathological in contemporary society, from feminism to homosexuality. Mothers caused
those pathologies by loving their children too much or too little. I say that
the contradiction is only apparent because the underlying logic of both
discourses is the same: the mother is the agent responsible for the emotional development of her children; therefore children's emotional problems result from pathological deviations in the mother's role. Too little
mother love impairs emotional development; too much leads to immature
individuals. Ribble had also made the logic of this clear: "Feeling, then,
is as fundamental in the life of the young infant as is food. As with food,
either starving or overfeeding is dangerous."[43]

Levy's views are a good example of the happy coexistence of those
seemingly contradictory discourses. While pointing out the disastrous
consequences of maternal deprivation, Levy also explored the equally
devastating effects of "maternal overprotection" in his 1943 book of that
title and other writings.[44] In his 1942 essay "Psychopathic Personality
and Crime," Levy argued that maternal rejection could lead to the "deprived psychopath" and maternal overprotection to the "indulged psychopath." One case history, fairly typical of those Levy reported, involved a
fourteen-year-old boy referred to him because of "stealing, truancy, and
incorrigibility." Twice this boy had been committed to a "protectory." The
psychiatrist's evaluation consisted of seven interviews with the patient,
one with the father (just released from prison), and one with the mother.
A social worker saw the mother on one occasion, and a worker in a co-
operating agency had interviewed her. The child's father was absent for
repeated prison sentences, and the mother was estranged from her own
family. But Levy said nothing about the father's possible effect on the
child or about the absence of other family members. He focused only
on the mother. His discovery: too much of a good thing can hurt you. In
Levy's assessment, this child suffered from maternal "spoiling."[45]

Levy's 1943 book *Maternal Overprotection* received overwhelmingly
positive reviews from numerous scholars, including psychologist Robert
Sears, sociologist Ruth Reed, anthropologist Ashley Montagu, and many
workers in children's agencies, pediatrics, and child psychiatry.[46] Their
reviews emphasized the soundness of Levy's methodological approach,

the scientific basis of the results, and their objectivity. The editors of the *Medical Record* felt confident enough to draw some policy implications: "After reading this book we believe that mothers should be given a test before they are permitted to raise their children in order to safeguard the health and welfare of the children, just as in many states we have a pre-marital test for syphilis before persons are permitted to marry and infect each other." Such a conclusion, though it might seem at first glance to be "radical and completely heartless," was deemed worth thinking about given that there were so many "ignorant mothers who wreck the children's lives through overlove."[47]

At the same time that Levy tried to convince the scientific community of the dangers of "overlove," a best-selling author of science *fiction* penned the clarion warning against doting mothers. In his 1942 book *Generation of Vipers*, Philip Wylie mounted a ferocious attack against "mom," the overbearing mother. By curtailing her children's independence, mom, asserted Wylie, was the source of the disrupted emotions of the two sexes in contemporary American society: "The mealy look of men today is the result of momism and so is the pinched and baffled fury in the eyes of womankind."[48] Mother was thus the source of the dysfunctional gender relations of the day, creating pale men and furious women, that is, emasculated sons and feminist daughters. *Generation of Vipers* became an instant best seller. In 1950 the American Library Association selected it as one of the major works of nonfiction of the twentieth century.[49]

The incredible success of Wylie's book and the widespread debate about his critique of mothers reflects how far child studies had succeeded in putting the mother at the center of wartime and then postwar social discussions about personality. Some psychoanalysts recognized Wylie as an ally. Karl Menninger told him, "There are some themes in your book which are very close to my own thinking," and he reported that the Menninger clinic would send copies of *Generation of Vipers* to all staff members who were away in the service.[50]

Only four years after the Levy and Wylie attacks, mom received another trenchant critique in Edward Strecker's 1946 book *Their Mothers' Sons: The Psychiatrist Examines an American Problem*. Unlike Wylie, Strecker had hefty scientific credentials, as noted after his name on the book cover: AM, MD, ScD, LittD, LLD. He was chair of the department of psychiatry at the University of Pennsylvania; consultant to the surgeon general of the army, air force, and navy; adviser to the secretary of war; and from 1943 to 1944 president of the American Psychiatric Association. Psychiatrist Eugene Meyer, chairman of the National Committee on Mental Hygiene, claimed in the book's foreword, "Dr. Strecker's background gives him the right to speak."[51]

Strecker's indictment of mothers was as strong as Wylie's, yet he tried hard to establish some intellectual distance between them. Whereas Wylie described mom "in too vindictive terms to satisfy a trained psychiatrist," Strecker said he would use "mom" as "merely a convenient verbal hook upon which to hang an indictment of the woman who had failed in the elementary mother function of weaning her offspring emotionally as well as physically." In truth, Strecker's views hardly differed from Wylie's. But by emphasizing his scientific credentials, he laid claim to the scientific objectivity and credibility of his analysis. In the foreword, Meyer emphasized that Strecker's position was strictly scientific, based on facts measured by reliable instruments.[52]

Strecker's point of departure was the alarming data on recruits during World War II, "the cold hard facts that 1,825,000 men were rejected for military service because of psychiatric disorders, that almost another 600,000 had been discharged from the Army alone for neuropsychiatric reasons ... and that 500,000 more attempted to evade the draft." Why did American men fail to make the grade? In Strecker's opinion, because they were "IMMATURE."[53] Having established the immaturity of men, Strecker inferred the cause without further ado: mom was at fault. According to Strecker, she failed to "untie the emotional apron string—the Silver Cord—which binds her children to her."[54]

Like Wylie, Strecker assumed that what he perceived as the emasculation of men resulted from the derailment of women's femininity. From the staggering number of American males unable or unwilling to perform as real men by waging war, Strecker concluded that American females were unable or unwilling to perform as real women by mothering in the right way. Strecker did not explain how he moved from the results to the cause. To establish his conclusions about "American youth," he had not carried out any research beyond the army tests. He had not collected any data on young men, women, American mothers, peer groups, or families. Nevertheless, he confidently identified mom as the cause of a variety of pathologies, from mental illness to alcoholism.

Mom allegedly led not only to psychopathologies, but also to sociopathologies. Expanding the scope of the underlying individual childhood determinism of his position, Strecker claimed that "the capacity to live democratically and constructively is acquired only in childhood."

Thus Strecker moved from the individual to the nation, from moms to "momarchies." If the character of a nation depends on the personality of its individual members, and if mothers determine individuals' emotional character, it was almost a logical conclusion that mothers are ultimately responsible for the fate of a nation.[55]

Despite the lack of research on the actual effects of mothering practices and maternal feelings on child development, the literature on momism and on maternal deprivation snowballed, influencing various scholarly and practical endeavors. In psychiatry, for example, it had profound consequences. In 1948 refugee Austrian psychoanalyst Frieda Fromm-Reichman introduced the concept of the schizophrenogenic mother. She argued that schizophrenic individuals had their early emotional needs warped by their mothers' rejection. In 1943 Leo Kanner diagnosed childhood autism as a condition that, among other things, prevented a child's right emotional development. He noted that the children he identified as suffering from autism all had educated mothers. Subsequently, a number of authors, including Kanner, selected maternal rejection as the cause of autism.[56]

Psychoanalysts continued to contribute to this discourse and gained prominence from the social interest in this area. Take the case of Erik Erikson (1902–1994), a German-born psychoanalyst who wrote *Childhood and Society,* one of the most influential texts. Many authors had argued that childhood is essential to understanding an adult's behavior. But there is a gap between infants' experiences and their adult behavior. Erikson filled this gap by postulating that human personality develops from one stage to the next, all the way from the embryonic infant self to the mature social self.[57]

The idea that personality, together with the body, develops through a series of steps, unfolding in a pattern that may be universal, was already part of child psychology. One of its best-known proponents was psychologist Arnold Gesell at Yale.[58] But Erikson now provided a specific timeline for emotional development. For Erikson, each stage in personality development occurs at a set time, when the individual confronts a specific crisis and undertakes certain tasks until the next stage is reached. In the first stage, the infant enters into the most significant relationship—with its mother—and develops the basic trust that is the foundation for healthy emotional development. The establishment of basic trust does not depend on the amount of food or parental care, but rather, in Erikson's words, "on the *quality* of the maternal relationship."[59] By claiming that later stages of personality growth cannot unfold without the earlier ones, Erikson turned the emotional support of mothers into the essential ingredient for individual development and, by extension, for social maturity.

In Erikson's view, maternal care determines not only the identity of an individual, but also the identity of a society. In his reflections on American identity, he argued that deep problems in American society resulted from a pervasive conflict between the personal and social ideals encouraged in the past and the ones demanded by the present historical circumstances. As a result, the ego's identity was now built on a combination of competing "dynamic polarities": migratory and sedentary, individualistic and standardized, competitive and cooperative, pious and freethinking, responsible and cynical, and so forth. These contradictions exposed American youth to a dangerous "emotional and political short circuit." To understand this situation, Erikson turned to the main source of emotional trouble: mom. For Erikson, mom was a composite, a "psychiatric syndrome" that one could use as a "yardstick." Though one could never see this fully implemented in a real person or in pure form, he found this "type" to be "of sufficient relevance for understanding the epidemiology of neurotic conflict."[60]

Erikson noted that the historical and sociological work necessary to explain the development of mom was yet to be done, but he believed he could identify the central factor in its genesis. Here he resorted to the explanatory workhorse of the American character: the frontier. According to Erikson, frontier life had required a set of maternal virtues that later became obsolete or problematic. Between the temptation to move on and the appeal of sedentary life was the cradle of the self-made ego, later reinforced by the machine age. But in the pursuit of adjustment to and mastery over the machine, American mothers (especially the middle class) found themselves standardizing and overadjusting children, who later were expected to personify that very virile individuality that in the past had been an outstanding American characteristic.

These historical developments had led to the distorted figures of mom and pop. In fact, Erikson claimed, "Momism is only misplaced paternalism," resulting when fathers "abdicated their dominant place in the family." Thus mom was "a victim, not a victor."[61] In a single explanatory framework, then, Erikson managed to present mother love as essential and dangerous while also simultaneously blaming and exculpating mothers.

In light of what we know about Erikson's personal struggle about his own identity, it is hard not to suspect that his experiences influenced his views about dislocated parental roles. He was born in Frankfurt in 1902 to Karla Abrahamsen, a young Danish Jewish woman. Although her son often asked her throughout his life, she never named his father. When Erik Abrahamsen was three years old, his mother married his pediatrician, Dr. Theodor Homberger, and Erik Abrahamsen became Erik Homberger. In a bold and revealing move, he also "fathered" himself by choosing a new name—Erik-son—in his petition for US citizenship in 1938.[62]

Erikson's professional success attests to the increasing acceptance of the research from the previous decade that he had built on and elaborated in *Childhood and Society*. An artist, he taught art at the experimental school for American students run by Dorothy Burlingham and Anna Freud in Vienna and underwent psychoanalysis. He did not have a bachelor's degree, but he obtained certificates in Montessori education and from the Vienna Psychoanalytic Society. In the United States, the wandering European artist successfully reinvented himself. Combining the ideas of current psychoanalysts with the flair of popular authors, Erikson gained recognition from the lay and scientific communities. His tremendous success must in part be explained by the timing of his writings on childhood and their social relevance.

Erikson's book integrated the significance of maternal love in the early years with the importance of separation from the mother later on, thus building on the insights of researchers working on maternal deprivation and on maternal overprotection. By the time Erikson published *Childhood and Society* in 1950, Benedek had already argued that children need to develop a basic sense of confidence or trust in the mother. Wylie and Strecker provided mom. Combining these notions with that of the frontier, Erikson provided a particular account of American identity. Moreover, he then developed a life plan that defines the normal and natural stages of universal emotional development. In this plan, the good mother enables the trust her infant needs to develop normally.

By highlighting the importance of maternal recognition in the first stage of emotional development, Erikson's 1950 book helped call attention to the importance of infancy. Just one year later, another psychoanalyst, John Bowlby, also wrote a synthetic work that would become a reference point in discussions of maternal deprivation.

John Bowlby: The Mother as the Psychic Organizer

Born in London into an upper-class British family, Edward John Mostyn Bowlby (1907–1990) was the heir to a baronetcy awarded to his father for his services as surgeon to the king. Bowlby received his education at Trinity College, Cambridge, spent a year working in a school for maladjusted children, and became interested in mental health. He obtained a medical degree at University College Hospital, London, and specialized in psychiatry at Maudsley Hospital. He also trained in psychoanalysis with Joan Rivière and Melanie Klein at the British Psychoanalytic Institute, gaining experience as a staff psychologist at the London Child Guidance Clinic and Training Center. After serving as a psychiatrist in the British army during World War II, in 1946 Bowlby became head of the children's department of the Tavistock Clinic in London, where he remained until his retirement.[63]

At the end of World War II, the United Nations commissioned a study about the needs of homeless children. G. Ronald Hargreaves, chief of the Mental Health Section of the World Health Organization (WHO) in Geneva, gave the task to Bowlby, whom he knew from his war work. Supported by the WHO, Bowlby spent several weeks in Europe and the United States visiting researchers working on a variety of children's issues, including delinquency among adolescents and separation from family during hospital stays, institutionalization, and foster care. In early 1950, on a trip to the United States, he visited Yale, the Putnam Center in Boston, the Orthogenic school in Chicago, and Washington, DC, among other places. He met many experts, including Spitz, Katherine Wolf, Levy, and others such as William Goldfarb, whose work he found particularly interesting and used extensively in his report.[64] In research carried out at the Foster Home Bureau of the New York Association for Jewish Children, Goldfarb found that many children who had been in institutional care could make only "partial love attachments" in their new homes. In other studies, Goldfarb compared institutionalized children with children in foster homes. He argued that the personality of children in institutions was "congealed at a level of extreme immaturity," and that they showed an "attitude of passivity and emotional apathy."[65] Putting together those results with the work of Anna Freud and Burlingham, Levy, Ribble, and Spitz, among other researchers, Bowlby wrote his report.

In 1951 the WHO published the report as *Maternal Care and Mental Health*.[66] As the title indicated, the main thesis was that maternal care is essential for a child's mental health. Bowlby first pointed out what he considered one of the most important developments of recent psychiatric research: "The steady growth of evidence that the quality of the parental care which a child receives in his earliest years is of vital importance for his future mental health." Since Bowlby focused only on the role of the mother, this thesis about parental care translated into the view that "*maternal care* in infancy and early childhood is essential for mental health." The father received the standard treatment of the time. Bowlby stated: "In what follows . . . little will be said of the father-child relation; his value as the economic and emotional support of the mother will be assumed."[67]

Bowlby presented his conclusion as based on the confluence of the results of various research teams all over the world, but he had already established the link between lack of maternal care and love and psychopathology well before he started his work for the WHO. In 1940 he published a review of "about 150 cases" of neurotic children who had been treated at the London Child Guidance Clinic, whom he had met a few times, and whose mothers he had interviewed. To understand the causation of neurosis from an "analytic angle," he had "ignored many aspects of the child's environment such as economic conditions, housing conditions, the school situation, diet and religious teaching," though "some psychiatrists" thought them to be "important."[68]

Bowlby focused only on the "*personal environment* of the child," which meant just the mother, since he did not meet or gather information about the fathers or the families. Instead, he identified twenty-two children who had experienced a break in the relationship with their mothers. Of those, "fourteen had become affectionless thieves and three had become schizophrenic." This group included children whose mothers had died, children in foster homes or in the care of relatives, and children hospitalized with major illnesses. But the basic cause of their mental problems, according to Bowlby, was the "dramatic interruptions of the child's emotional development" that resulted from the "broken mother-child relation." Moreover, he did not consider the possibility that the breakdown in the mother-child relationship could be the result, rather than the cause, of the children's mental problems.[69]

Then Bowlby turned to a second group of children who had "never suffered any obvious psychological trauma" and who had "remained in a relatively stable home." Though looked after by their mothers and "well cared for according to ordinary standards," they had "developed into neurotic children with great anxiety and guilt and abnormally strong sexual and aggressive impulses." After investigating the causes of their troubles, he claimed that one factor stood out: "the personality of the mother and her emotional attitude towards the child." In some cases the mother "had strong unconscious hostility towards her child," which could be observed in "unnecessary deprivations and frustrations, in impatience over naughti-

FIGURE 1.4. "Just child's play." Photograph of child therapy session with Dr. Bowlby used to illustrate an article, ca. 1950. John Bowlby Papers. PP/BOW/L.2 (Image Loo25169). Wellcome Library, London.

ness, in odd words of bad temper, in a lack of the sympathy and understanding which the usual loving mother intuitively has."[70]

In conclusion, lack of maternal care was conspicuous as the main factor leading to affectionless criminals, to psychopaths, and to neurotic, aggressive, oversexed, and anxious individuals. Bowlby reached similar conclusions in a subsequent study, "Forty-Four Juvenile Thieves: Their Characters and Home-Life," where he identified a *"prolonged separation"* from the mother as the "outstanding cause" of the children's emotionally unstable character and delinquent behavior.[71]

Note that what Bowlby included under maternal care was not simply the mother's attentiveness to her child's physical or even emotional needs. The presence of the mother is necessary but not sufficient for mental health; he considered the *emotional quality* of the mothering just as important. The sympathetic love that a mother "intuitively" feels for her child and the mother's "unconscious" feelings were also crucial for normal development. The lack of such a natural feeling could be expressed

in myriad ways, from impatience to unruliness, from unnecessary deprivations to too much spoiling.[72] Thus, good mothering required not just love,
but love with certain characteristics. According to Bowlby, the most important thing is for the child not only to feel mother love, but also to sense
that the mother is happy with that love. Bowlby repeated this point a decade later in the WHO report: "Just as the baby needs to feel that he belongs to his mother, the mother needs to feel that she belongs to her child,
and it is only when she has the satisfaction of this feeling that it is easy
for her to devote herself to him." According to Bowlby, psychiatrists and
child guidance workers believed that "the infant and young child should
experience a warm, intimate, and continuous relationship with his mother
(or mother-substitute), in which both find satisfaction and enjoyment."[73]

Following the emphasis on the natural basis of the need for mother
love, Bowlby also adopted the analogy between love and vitamins, which
had been introduced by David Levy and used by Margaret Ribble, René
Spitz, Anna Freud, and other psychoanalysts. Bowlby put forward his own
variation on the theme: "mother-love in infancy and childhood is as important for mental health as are vitamins and proteins for physical health."[74]

To emphasize the natural basis of the mother's role, Bowlby resorted
to a biological process, embryological development. German zoologist
Hans Spemann (1869–1941) had introduced the concept of the "organizer" to explain the results found by one of his students, who transplanted
the dorsal lip of a new gastrula into a host embryo and found that it in-
duced the host tissues to develop a secondary axis, including a central
nervous system. Spemann called the dorsal lip the "organizer." Spemann
understood embryonic development as a sequence of orderly inductive
developments, guided by the primary induction of the organizer. Bowlby
did not refer to Spemann, but he adopted the notion of the organizer and
established a parallelism between physical and psychic development. Just
as embryologists had proposed the existence of "organizers" to guide the
growth of the embryo, Bowlby argued that "if mental development is to
proceed smoothly, it would appear to be necessary for the undifferenti-
ated psyche to be exposed during certain critical periods to the influence
of the psychic organizer—the mother."[75] This analogy with embryology
highlighted the key position of mother love as the primary inducer of an
orderly emotional development. Like numerous researchers, including
many of those mentioned above, Bowlby emphasized the natural basis
of the mother-infant dyad, said that mother love is a nourishing material,
and established an analogy between emotional development and physical
development.

But soon Bowlby went further than other authors, claiming that important animal research supported his views. Although he believed his "main proposition" about the need for maternal love could be "regarded as established," he recognized that "knowledge of details" remained "deplorably small," in part because of the difficulty of studying children. However, further evidence for his thesis could be found, he claimed, "if we use animals." In earlier writings Bowlby had looked at research with animals for analogies and evidence.[76] He could not include much in the first WHO report, but the situation changed right after its publication. The reason? Bowlby had encountered ethologist Konrad Lorenz's work on animal behavior.

In 1953 Bowlby published a shortened version of his WHO report, *Child Care and the Growth of Love*, which was hugely successful. It was reprinted six times in the following ten years, was translated into fourteen languages, and sold over 400,000 copies in the English paperback edition.[77] For this version, Bowlby made hardly any changes, but he considered his discovery of ethology so important that he added several pages to the first chapter. This addition has not been noted in the literature on Bowlby, but it marked a key transitional moment in his thinking, and it would play a major part in the acceptance of his views. Bowlby noted that the idea that "emotional experiences at certain very early and special stages of mental life may have very vital and long-lasting effects" had been "shown to be true of birds and dogs" in studies by some "European biologists." Although Bowlby did not name anybody in the text, the reference he made in his "List of Authorities referred to but not named" was to Lorenz. Bowlby did not explain how the "behavior" of dogs and birds is relevant to understanding the emotions of human infants. Yet he used this reference to assert that his own theories, "far from being in themselves improbable, are in strict agreement with what biological science has shown to be true of both bodily and mental growth."[78] Bowlby's appeal to animal studies would prove crucial for the success of his ideas.

The authors working on maternal deprivation presented mother love not merely as something a child desires or benefits from, but as an instinctual need. As Ribble put it, "Babies have the right to develop to the full the natural resources of life energy within them called instinct."[79] Yet though many authors called for studying the biological basis of the mother-child unit and appealed to its instinctual basis, none of them clarified the role of biology in shaping the infant's psyche. By appealing to the work of biologists, Bowlby initiated an interdisciplinary move that would become highly fruitful—and highly contested.

Conclusion

During and after World War II, the fear of a dehumanized, mechanized, and destructive world had led to an emphasis on the natural basis of emotions and behavior. To a certain extent, this project of humanizing the modern world depended on reaffirming the natural aspects of human nature and, more specifically, on reasserting the importance of one of the most natural relationships: that of infants with their mothers.

Seen as a determinant agent in childhood and social well-being, the mother acquired power of enormous proportions. This power gave rise to an apparently contradictory discourse of mother blame: mothers loved their children either too much or too little. Popular writer Philip Wylie and psychiatrist Edward Strecker were the most visible critics of mothers who allegedly loved their children too much and thus curtailed their independence. Psychoanalyst Erik Erikson extended the power of mom from her children to the whole of society. In their writings, since mothers determine their children's personality and childhood experiences determine the personality of adults, mothers are ultimately responsible for the character of a nation.

Although the discourse on momism remained alive and well into the 1950s, during the late 1940s most research in psychoanalysis focused on early infancy and the perils of maternal deprivation. The work of Anna Freud, Margaret Ribble, René Spitz, David Levy, Therese Benedek, Erik Erikson, and John Bowlby was crucial in this respect. This literature on the mother-infant dyad fostered an increasing interest in children's instinctual needs. Many of the researchers working on maternal deprivation suggested that the infant's need for mother love is biological.

John Bowlby's report on maternal care and mental health became the synthetic work that would turn maternal deprivation into a major concern of the 1950s. Bowlby went beyond other researchers by claiming that his view of the mother as the psychic organizer was supported by research on animal behavior. Perhaps, as many child analysts suggested, biology could help psychology understand the nature of instincts. In the next chapter, I examine the work of the animal researchers Konrad Lorenz and Niko Tinbergen, whose ideas on social behavior would transform Bowlby's views on child development.

CHAPTER TWO

The Study of Instincts

Introduction

In his 1953 book *Child Care and the Growth of Love*, John Bowlby asserted: "The theories put forward in this book, . . . are in strict agreement with what biological science has shown to be true of both bodily and mental growth." He referred to the "European biologists" who had shown in birds and dogs that "emotional experiences at certain very early and special stages of mental life may have very vital and long-lasting effects." The European biologists he was referring to were the Austrian Konrad Lorenz and the Dutchman Nikolaas Tinbergen, two students of animal behavior who aimed to institute the biological study of instincts as an independent branch of research: ethology.[1]

Several scholars have illuminated different aspects of the history of ethology. Richard Burkhardt, John Durant, Theodora Kalikow, Robert Richards, and others have examined central conceptual and institutional developments in the field. Klaus Taschwer and Benedikt Föger's biography of Lorenz and Hans Kruuk's biography of Tinbergen offer fascinating accounts of these two complex men. Together with D. R. Röell's study of ethology in the Netherlands and Burkhardt's comprehensive book about research on animal behavior, this literature provides a thorough account of the development and reception of ethology.[2] Here I expand on those aspects of ethology that influenced ideas about infants' needs and maternal care and focus on Lorenz's influence in child studies, a topic that has received less scholarly attention.

Ethology, and Lorenz's work on imprinting in particular, became fundamental to the construction of mother love and love for mother as biological instincts. To show this, I first introduce Lorenz's and Tinbergen's project to explain the biological basis of animal and human behavior. They believed that human as well as animal behavior is a matter of instincts. I then examine more closely Lorenz's conception of instincts and imprinting. Next I turn to Lorenz's role in discussions about maternal care and infants' needs. After psychiatrists like Bowlby showed interest in his work, Lorenz increasingly extrapolated his ideas about imprinting in birds to the human realm.

Ethology: Lorenz and Tinbergen Search for the Biological Basis of Behavior

Konrad Lorenz (1903–1989) was the son of a world-famous orthopedic surgeon and professor at the University of Vienna. Lorenz arrived late in his parents' marriage, eighteen years after his only brother. He grew up in Altenberg, a small Austrian town near Vienna, in a mansion whose extensive grounds, and the surrounding forest near the Danube, served as his playground and, later, as his observational and experimental site.[3] In this wealthy, educated household, his devoted parents indulged his passion for animals. Lorenz raised dogs, fish, and a variety of birds with the help of Margarethe (Gretl) Gebhardt, a neighbor's daughter with whom he shared games and animals beginning in childhood. Following his father's desires, Lorenz studied medicine at the University of Vienna, where he earned his doctorate in 1928. In 1933 he also obtained a doctorate in zoology there. A few years earlier, in 1927, he had married Gretl, who became a medical doctor and a lifelong supporter of her husband's projects.

Lorenz was interested both in animal and human behavior from early on, as can be seen in his training in comparative anatomy, ornithology, and human psychology. In medical school, Lorenz studied with the renowned comparative anatomist Ferdinand Hochstetter. Under his tutelage, Lorenz learned to identify homologous anatomical structures and use them to trace common ancestries. Later, Lorenz used instinctive behaviors to reconstruct evolutionary phylogenies. In ornithology he received the approval and encouragement of the influential German ornithologist Erwin

Stresemann. In addition, the assistant director of the Berlin zoo Oskar Heinroth, who studied instinctive behavior patterns in birds, also mentored Lorenz. In human psychology, Lorenz was a student of Karl Bühler, director of the Psychological Institute at the University of Vienna and a world leader in several areas of psychology, including Gestalt perception and child development. Bühler had also written about instincts and, in a successful book that went through several editions, reviewed the various schools of psychology, including Gestalt psychology, behaviorism, and psychoanalysis. Lorenz did one of his auxiliary fields for his PhD exams with Bühler and regularly attended his seminars.[4]

From Bühler, Lorenz learned about the major discussions in human psychology, such as the debate between psychologists William McDougall and John Watson on the role of instincts in human conduct. Whereas McDougall defended the notion that humans were moved by several key instincts and emotions, Watson argued that they were born with very few innate predispositions. Lorenz soon concluded that much work remained to be done to prove the importance of instincts in human social behavior. Convinced of the scientific and social significance of his goals, Lorenz felt ready for that task. Observing birds, he wondered what his studies implied for human behavior. As he put it in a 1931 letter to Heinroth: "Who knows what will become of today's human psychology if one can only know what is instinctive behavior and what is rational behavior in humans? Who knows how human morals with their drives and inhibitions would look if one could analyze them like the social drives and inhibitions of a jackdaw."[5]

Only a few years later, Lorenz presented a unitary framework for explaining animal and human social behavior. Between 1927 and 1935, he published detailed observations of the behavior of half-tamed jackdaws, ravens, and night herons, as well as his views about instincts.[6] Then in a 1935 paper, "Companions as Factors in the Bird's Environment: The Conspecific as the Eliciting Factor for Social Behaviour Patterns," he put forward a general framework for understanding social behavior that he would defend with only minor modifications for the rest of his career. In this massive essay, Lorenz included observations on almost thirty types of birds and described their social behavior as a set of instinctive responses that had been built by natural selection because of their survival value.[7]

Lorenz posited the existence of innate templates or schemas (later called innate releasing mechanisms) that, when activated by internal and external releasers, lead the bird to perform certain instinctive behaviors.

He elaborated on the role of various fellow members of the species, the "companions," that act as releasers of a bird's social behavior: the parental companion, the infant companion, the sexual companion, the social companion, and the sibling companion. According to Lorenz, the bird's reactions to the specific stimuli from each of these companions form a functional system. For each of these relationships, there is an innate releasing mechanism, some type of conduct or image in the companion that functions as releaser, and an instinctive pattern of behavior that is "released" in an automatic and uniform manner.[8]

In this paper Lorenz asserted that "social behavior patterns in particular are to a large part governed by instincts even in the highest animals." He concluded with a call to arms, urging researchers to "recognize that the instinct, governed by its own laws and fundamentally differing from other types of behavior, is also to be found in human beings" and encouraging them to investigate human instincts.[9] At this point, however, Lorenz continued studying birds with renewed energy after meeting Niko Tinbergen, a young researcher with whom he had been corresponding.

Nikolaas Tinbergen was born in The Hague in 1907, the son of a grammar school teacher. He and his four siblings were raised in an austere household by parents who had a strong sense of public duty. Throughout his youth, Tinbergen was more interested in the outdoors and in field hockey than in intellectual matters. After a trip to an ornithological field station in Germany, he decided to turn his interest in wildlife into a career in field zoology.[10]

Tinbergen obtained his PhD in biology at Leiden University in 1932 with a dissertation on the homing behavior of the digger wasp *Philanthus*, based on field experiments he had done in the sand dunes area of Hulshorst. After he married Elisabeth (Lies) Rutten, they spent one year in Greenland with a meteorological expedition. Tinbergen then returned to his post as assistant in the Department of Zoology at Leiden University. With his students, Tinbergen continued his experiments on the hunting and homing behavior of wasps, added observational work on the parental behavior of hobbies (birds), and carried out experiments on the territorial behavior of sticklebacks (fish). When he met Lorenz, he was ready for a theoretical framework that could help make sense of those diverse observations and experiments.

At first sight Lorenz and Tinbergen seemed an unlikely team because they had very different personalities, but they immediately found common ground in their shared interests. Flamboyant and self-assured, Lorenz reveled in being the center of attention at scientific and social gatherings. He was an animated speaker who, much to the delight of his audiences, readily went down on all fours to imitate a variety of animal sounds and gestures. Convinced of the scientific and social importance of his work, he presented his views with messianic zeal. Tinbergen, on the other hand, battled with self-doubt throughout his life and often questioned the value of his contributions to science and society. But they both enjoyed being outdoors and were drawn to physical activities. Above all, they shared a deep love for nature. Both men had spent their childhood and youth in the open air, observing animals, and they continued this throughout their lives.

When Tinbergen spent three and a half months with Lorenz at his Altenberg home and research site in 1937, they carried out observations and experiments on a variety of birds. Among the best known are their experiments to test young turkeys' reaction to simulated predators and to analyze the egg-retrieval movements of greylag geese. In the latter, they filmed a goose rolling a stray egg back to her nest with her beak. Once the goose started, she would automatically continue her motion toward the nest even when the investigators replaced the egg with objects like a huge egg or a cube, and even when they took the egg away altogether.[11]

Lorenz and Tinbergen believed that their individual research skills complemented each other. Lorenz was impressed by Tinbergen's ability to design experiments, a good counterpart to his own interest in observation. Tinbergen admired Lorenz for his theoretical ideas, while Lorenz realized that his theories would benefit from the evidence provided by Tinbergen's work. So joining forces would benefit them both. From then on, although they developed most of their research separately, they conceived of their work as part of a common enterprise.

The two men decided to build and promote a modern biological science of animal behavior—ethology—focused on the study of instincts. This was not an entirely new endeavor. As Richard Burkhardt has shown, many of the elements of their approach to studying instincts were already present in the work of researchers of the previous generation. What distinguished Lorenz and Tinbergen was their concerted effort to make the study of instinct a distinct and central branch of biology.[12] Together they defended the notion that observation should be the key method for studying animal behavior, and they encouraged the use of photography in research.[13]

Tinbergen promoted their ethological program during a 1938 trip to the United States, where he delivered a series of lectures at the American Museum of Natural History and the Linnaean Society of New York and participated in other scientific meetings. Meanwhile Lorenz remained in Austria, continuing his research, pursuing funding, and trying to secure a permanent position. The prospects for ethology looked promising.

But World War II rudely interfered, putting a stop to their efforts to institutionalize ethology as an independent branch of biology and separating them geographically as well as politically.

In 1938, soon after Germany annexed Austria, Lorenz joined the Nazi party and sought to get ahead in a regime that he thought would support his scientific interests, including human psychology. He tried to obtain the chair in human psychology at the University of Vienna left vacant by the Nazis' dismissal of his former teacher Karl Bühler, "for political and world-view reasons." His wife, the respected psychologist Charlotte Bühler, had one Jewish parent and was dismissed on racial grounds.[14] They both emigrated and eventually found jobs in the United States, although Karl Bühler never again worked at the prominent level he had attained in Europe. Lorenz did not obtain his position, but two years later he was awarded a professorship of psychology at Albertus University in Königsberg, in East Prussia. Barely a year later, however, he was drafted into the German army.

After a brief stint as a motorcycle-riding instructor, Lorenz spent two years (from summer 1942 to spring 1944) as a psychiatrist in Poznan, Poland. From the scarce documents available, historians have been unable to determine whether his work as an assistant to race psychologist Rudolf Hippius contributed to Nazi activities. Later Lorenz denied his membership in the Nazi party and said he never suspected the Nazis had mass murder in mind.[15]

During this period, in addition to articles on bird behavior and the concept of instinct, Lorenz wrote on the consequences of interfering with instinctive behavior in animals and humans. Comparing wild and domesticated animals, Lorenz defended the superiority of the wild forms. He pointed to the short extremities, fat belly, and early sexual maturation and promiscuity of females in some domestic breeds as evidence of degeneration from their wild state. Lorenz believed that civilization in humans, like domestication in animals, led to degeneration owing to the deterioration of innate releasing mechanisms. In humans, he argued, civilization led to physical degeneration as well as a decline in innate moral and aesthetic capacities.[16]

Presenting domestication in animals and civilization in humans as analogous helped Lorenz to justify extrapolating animal research to the human realm and also to foreground the social relevance of his own expertise. Note the elaboration of this chain of reasoning in the following text Lorenz published in a 1940 newspaper:

> Fighting spirit and motherly love, the characteristics that are necessary for the preservation of the species, are being lost not only in animals, but in humans as well, through the process of civilization. This is why one may draw comparisons between the two realms without further ado. Race politics knows that the continuous ups and downs, the flowering and decline of cultures arise when the victorious people become complacent. Today, the biologist researches consciously and with scientific precision the causes of these phenomena.[17]

Here Lorenz established the social significance of his expertise by arguing that biologists could help to improve the race, or at least to avoid its further degeneration. In turn, the urgent need for biological knowledge justified the extrapolation from animals to humans. By focusing on a major Nazi concern, race purity, Lorenz defended his views about the perils of civilization for the degeneration of the human race while advancing the importance of his scientific work.

During the Nazi period in Germany and Austria, Lorenz took advantage of a political climate favorable to these ideas by presenting his views in the rhetoric prevalent at the time. In several writings dating from the early 1940s, he couched his concerns about social decadence in the language of racial degeneration and used medical metaphors to cast the alleged deterioration of human instincts as a cancer of the social body. Although he had not done any research on the behavior of humans, in 1943, eight years after his call to investigate instincts in humans in the "Companions" piece, he published another massive paper, "The Innate Forms of Possible Experience," in which he summarized his views on the major areas of his thinking: animal behavior, evolutionary epistemology, the instinctual nature of human behavior, and the negative impact of civilization on human instincts. He postulated the existence of innate releasing schemas of behavioral systems that shape human experiences, including those related to aesthetic and ethical appreciation.[18]

This essay, written while he was a military psychologist in occupied Poland, was Lorenz's last publication before the end of the war. In April 1944 he was sent to the front as a physician and almost immediately went missing. The Russians sent him to a prisoner-of-war camp in Yerevan, Armenia. By February 1946, his wife heard about his situation. In February 1948, he returned to Austria.

Back home and in need of income, Lorenz published a book for the general public about his experiences raising and living with animals. In German, the title was *Er redete mit dem Vieh, den Vögeln und den Fischen* (He spoke with the beasts, the birds and the fish). It was translated into English as *King Solomon's Ring: New Light on Animal Ways*. This was not a scientific book on animal behavior, but a collection of anecdotes that showcased Lorenz's knowledge of birds and other animals as well as his enormous talent as a storyteller. He described animal behavior in a dramatic narrative consisting of major episodes of human interest such as falling in love, marriage, and fighting. This was embellished with tales of how his family life was deeply intertwined with the animals he was studying and illustrated with his own charming drawings. The book became a best seller and made Lorenz a household name internationally.[19]

Reunited with his fellow ethologists in Austria and Germany, Lorenz resumed his research on animal behavior. He obtained some funds from the Austrian Academy of Sciences to continue his work in Altenberg. In 1950 support from a German foundation, the Max Planck Society, enabled Lorenz, his family, some collaborators, and many animals to move from his deteriorating mansion to the castle of a German aristocrat in Buldern, in northern Germany, where he continued his research. In 1958 the Max Planck Society provided him with a permanent research setting. Lorenz and the physiologist Erich von Holst were made directors of the Max Planck Institute in Behavioral Physiology in Seewiesen, in southern Germany.

During the war, Tinbergen also went through dramatic experiences that affected him profoundly. Together with 80 percent of the faculty of his university, Tinbergen protested the dismissal of Jews and the Nazification of Dutch universities. He was detained and spent the war in a prison camp. With the paper he was allowed to use, he wrote a weekly letter to his children, telling them animal stories that he illustrated and would later publish in books for children. After his liberation, Tinbergen returned to teach at Leiden University in difficult postwar circumstances. When conditions improved, the university created a chair of experimental zoology for him. But Tinbergen, who was interested in expanding ethology outside the Continent, sent out feelers to colleagues in England and the United States. In late 1948 he accepted an offer to be a lecturer in animal behavior at Oxford University, where he later became a professor of zoology.

Lorenz and Tinbergen reconnected at a 1949 symposium, "Physiological Mechanisms in Animal Behavior," organized by the British Society for Experimental Biology in Cambridge. Although still bitter about the war and disappointed by Lorenz's support for the Nazis, Tinbergen believed that reestablishing international connections with German scientists was necessary to heal the wounds of the past. Seeing Lorenz as the one who had provided the main conceptual framework for ethology, he decided to renew their relationship for the sake of the field.

Energized by the opportunity to launch his program again at the Cambridge symposium, Lorenz staked his claim to the independent status of ethology as a science. First he advanced the object of study: "The distinct and particulate physiological process whose discovery may be identified with the origin of comparative ethology as an independent branch of science is represented by a certain type of innate, genetically determined behaviour patterns." As for method, Lorenz argued that "unprejudiced observation" was the appropriate way to study genetically determined behavior patterns. In his view, ethologists should start by developing "a morphology of behaviour" by "a thorough observation and description of *all* the behaviour patterns at the disposal of the species." The systems of behavioral patterns uncovered by ethology would then reveal natural laws of universal character.[20]

Drawing extensively on the conceptual apparatus he had developed in the 1930s, Lorenz again proposed a unitary framework for animal and human behavior. This implied considering human psychology from the standpoint of the central object of ethology, the instincts. He claimed that "the laws we have found in animal behaviour find an enormously important application to the special phenomena of human psychology." However, he did not specify which laws or which applications he was referring to.[21]

Tinbergen presented the ethological program in his 1951 book *The Study of Instinct*, which became the foundational text of ethology. As the "scientific study of behaviour," ethology first selected "a specific object, a special group of phenomena: innate behaviour."[22] Tinbergen also presented Lorenz's views on human instincts, arguing that Lorenz's writings in this area had already provided substantive evidence for the instinctual nature of human social behavior.

In sum, Lorenz and Tinbergen aimed to create an independent science of ethology that would explain social behavior by focusing on the study of instincts in animals and humans. But what, exactly, were the main objects of this science, the instincts? In the next section I focus on Lorenz's views about instincts, for two reasons: During the early years of ethology, Lorenz was the main theorist of the notion of instinct. And Bowlby and other child analysts borrowed the concept from his writings.

The Nature of Instincts

Lorenz and Tinbergen portrayed social behavior as guided or even determined by instincts. But what were instincts? And how could an investigator identify instinctive behavior? These were key questions for their ethological program.

Lorenz originally thought of instincts as reflexes, or chains of reflexes, set off by external stimuli, but he abandoned this view under the influence of the German physiologist Erich von Holst and the American biologist and animal researcher Wallace Craig. In 1937 Lorenz met Holst, who had shown that automatic behaviors generally cataloged as reflexes, such as creeping movements in earthworms, were caused by internal stimuli. Richard Burkhardt has shown that Lorenz was also heavily influenced by his correspondence with Craig.[23] Noting that animals are restless before performing an instinctive act but usually relaxed afterward, Craig argued that animals desire to perform their innate behaviors. In addition, the animal sometimes performs the action in the absence of the relevant environment or stimulus ("vacuum response").[24]

In Lorenz's new vision of animal behavior, the organism's central nervous system produces a series of internal stimuli that lead to an appetitive behavior, expressed in restlessness and searching. Impulses build up inside the organism. Then they are unleashed by an innate releasing mechanism (IRM) and by social releasers (SR), which "open the doors" to the organism's appetitive actions. Usually the IRM discharges in response to an external stimulus, but it can also explode "in vacuo." On some occasions "displacement activities" substitute for the organism's normal, natural, or instinctive actions.[25] For Lorenz every fixed-action pattern, a stereotyped action performed by all members of a species, has its own independent drive system. If a fixed-action pattern is not carried out, its "action-specific energy" builds up until it is eventually released. The intensity of a behavioral response depends on the amount of accumulated action-specific energy and the strength of the external stimulus.[26]

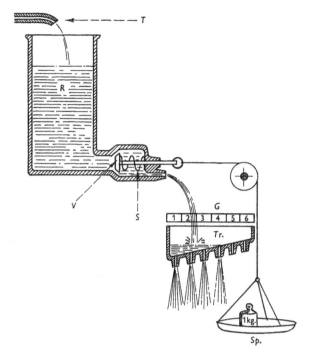

FIGURE 2.1. Konrad Lorenz's psychohydraulic model of instinctive action. From Konrad Lorenz, "The Comparative Method in Studying Innate Behaviour Patterns," *Symposia of the Society for Experimental Biology* 4 (1950): 256.

Lorenz devised a psychohydraulic model to explain how instincts work. Although he introduced a new vocabulary to refer to the components in the model, he relied on concepts of drive, energy, and release prevalent at the time in physiology, psychology, and psychoanalysis.[27] Like many of those models, Lorenz's motivational model is analogous to a hydraulic reservoir. He presented the following diagram: The stream flowing from the tap represents the internal accumulation of action-specific energy. The spring represents the central inhibition that must be overcome. The valve is the internal releasing mechanism. The liquid spouting from the lower tap represents the instinctive movement (see fig. 2.1).

The objects of study in ethology—instincts—sometimes went by different names: drive activities, instinctive movements, instinctual behavior patterns, fixed-action patterns, fixed motor patterns, innate behavior, innate behavior patterns, "genetically determined behavior patterns," or simply instincts. At the 1949 meeting in Cambridge, Lorenz had proposed "*endogenous movements*" to unify the diversity of concepts, but this term never acquired widespread acceptance, and Lorenz himself continued to use a variety of names.[28]

Many students of animal behavior also used these terms interchangeably, but it is important to note some important differences among them. A fixed-action pattern, one of the most common names in the ethological literature, refers to a pattern of behavior that is a stereotyped action performed by all members of a species. Whether animals perform an action in a stereotyped way can be ascertained by observation. But the other names all refer to something that cannot be determined by observing a behavior, namely, its innateness or instinctual nature. Tinbergen and Lorenz were interested in innate fixed-action patterns. But how could innate or instinctive behavior be identified?

According to Lorenz, instinctive actions share a series of features: they are species-specific, carried out by all members of a species in a stereotyped way; they always follow a given stimulus; and they continue until their consummation. That is, they are "machinelike" in character, an automatic sequence that, once set in motion, continues inexorably until its programmed completion. These behaviors are also satiable. Consequently, they are less likely to take place after several repetitions. But the action is sometimes performed "in vacuo" when the appropriate stimulus is not present.[29]

Further, the central characteristic of these innate behavior patterns is that they cannot be modified by training. Contrary to other theorists of instinct such as Darwin and William James, Lorenz posited a radical separation between instinctive and learned behaviors. In a given behavior, there may be a conglomerate of both types, what he called an "intercalation" of the innate and the learned (*Instinkt-Dressurverschränkung*), but the innate and learned components can be separated. In addition, he thought that an instinct could be treated "*as an organ*, whose individual range of variation can be neglected in the general biological description of a species." In Lorenz's account, the instinctive is uniform across the species.[30]

Although many of the characteristics of instinctive behavior can be observed, Lorenz argued that to be sure a behavior is innate, one needs to perform deprivation, or "Kaspar Hauser," experiments.

One day in May 1828, a disoriented young man was found in a town square in Nuremberg, Germany. He could not talk, walked with difficulty, and was afraid of many objects and animals. The authorities concluded that Kaspar Hauser, as he later called himself, had been locked in a cellar from infancy and had lived in isolation from the external world. His case, like those of other severely deprived children and "wolf children," or wild children, fascinated scientists because humans raised in isolation constitute a sort of experiment that can illuminate the nature-nurture controversy. The reasoning goes like this: if a behavior or trait that cannot have been influenced by training appears in an individual, then it must be inborn or instinctive.

The justification for using isolation experiments to find animals' instincts is based on a similar intuition. Lorenz believed a behavior that appears in an animal raised in isolation has to be innate, since it is not the result of imitation, practice, exercise, conditioning, or any other type of learning. Lorenz's mentor Oskar Heinroth and his wife, Magdalena, had engaged in an extensive program of such experiments in the early decades of the century to ascertain what was innate and what was learned in various species of birds from central Europe.[31] Following their insight, Lorenz believed that through observations and isolation experiments, researchers could identify an animal's instincts.[32] This was a key point, since he claimed that most social behavior is instinctual or innate.

Although Lorenz said that many social patterns of behavior are innate, he also claimed that the object that will release those patterns is not innate, at least in some species. Instead, it is acquired through imprinting.

Imprinting

As I mentioned before, in his 1935 paper "Companions as Factors in the Bird's Environment," Lorenz introduced the bird's companions as factors eliciting its instinctive behavior. In this paper, Lorenz also examined imprinting, which his mentor Heinroth and others before him had already observed in some types of birds and which was well known to farmers and breeders.

Imprinting is the process whereby some species of birds attach to the first moving object they see after hatching. Usually that object is their mother, and through that process the bird comes to recognize its conspecifics and develops the social responses of its species. The process of imprinting is very important for these species because, although their social behavior patterns are innate, the birds do not recognize the members of their species until they are imprinted. Lorenz had observed this phenomenon in jackdaws, geese, and many of his hand-reared birds. Lorenz elaborated on the significance of imprinting, and described its main characteristics:

FIGURE 2.2. Konrad Lorenz followed by birds imprinted on him. Courtesy of Konrad Lorenz Archive, Altenberg, Austria.

> I have described this behavior of the Greylag gosling because it provides a virtually classic example of the manner in which *a single experience* imprints the relevant object of the infantile instinctive behavior patterns in a young bird which does *not* recognize this object instinctively. *This object can only be imprinted during a quite definite period in the bird's life.* A further important feature is the fact that the Greylag gosling obviously "expects" this experience during a receptive period, i.e., *there is an innate drive to fill this gap in the instinctive framework.*[33]

For Lorenz, imprinting has clear characteristics. First, it takes place during an early critical period. Afterward the brain, like hardened wax, cannot be molded. Second, the bird has an innate urge to acquire the imprint of its species and will do so through a single impression, without conditioning, trial and error, or any learning period. As Lorenz put it, imprinting *"has nothing to do with learning."*[34] Third, it is irreversible.

By fixing the object of the bird's reaction in infancy, imprinting also determines future social and emotional reactions. If the infant bird is not imprinted on a member of its own species, it will still develop the instinctive social and sexual behavior patterns characteristic of its species, but it will direct them toward the wrong object. For example, some of the birds imprinted on Lorenz courted him later. Imprinting thus has irreversible consequences for the animal's behavioral development.[35]

In normal circumstances, the mother provides the image of the right species, filling the "gap" in her infant's instinctual framework. The mother thus enables the baby bird to release its instinctive social behavior toward the appropriate objects. In that sense the mother-infant relationship is essential for the bird's adequate development. One could say the baby bird has a built-in, instinctive need for the mother. Furthermore, the mother has also been designed by evolution to provide the appropriate responses to her infant.

The mother's behavior toward her offspring is also innate, according to Lorenz. In his view, one should not talk about the parental instinct in the singular because parental care includes many small innate components. In birds, it includes building a nest, then feeding and protecting the chicks. Lorenz considered these behaviors innate, especially those involved in maternal care. Discussing the innate schema of the infant companion, Lorenz noted that in most cases parents recognize their progeny instinctively. In addition, the characteristics that release parental conduct cannot be acquired through learning, since "the adult bird's own offspring are of course the first freshly-hatched conspecifics which it sees, and yet it must react to this first encounter with the entire repertoire of parental behavior operating to preserve the species."[36]

In his writings on humans and other animals, Lorenz usually talked about parental behavior or parental care or parental love, but most of his examples of those behaviors and emotions came from the female of the species. He argued that the chick's innate responses form a functional whole with the responses of the mother. Invoking his major test for innateness, Lorenz noted that the fact that the mother engages in maternal care toward her first infant without previous training proves its instinctive character. Furthermore, Lorenz claimed that maternal behavior is automatic and independent of the offspring's behavior; this too confirmed its instinctive nature. For example, he reported that a *Cairina* mother would "rescue" a mallard duckling from the experimenter's hands, even though minutes later she would bite and kill it when it tried to mix with her own chicks. In a sense, the mother cannot refrain from helping the infant in the first place. Her instincts impel her to act. For Lorenz, the "automatic nature of these parental care responses" proves that "the unitary treatment of the offspring is thus determined within the instinctive framework of the adult bird and not in the rôle that the infant plays in its environment."[37] That is, the actions involved in caring for the offspring are predetermined in the instinctive framework of the parents.

In sum, Lorenz postulated the existence in some species of birds of an innate need for imprinting—a process that in normal circumstances leads an infant to attach to its mother. He also postulated an innate mechanism that leads a mother to look after her infant. In the case of maternal care, the mother's actions are predetermined within her instinctive framework. When Lorenz talked about innate behavior and its releasers, he often employed the metaphor of a lock and key. For any particular lock, "the form of the key-bit is predetermined."[38] The same holds for the behavior within the mother-infant dyad. The preservation of the mother-infant behavioral system, understood as an interlocking system of instincts, is essential for adequate infant development. But could these findings be generalized to humans?

For humans, Lorenz postulated the existence of an innate releasing schema toward babies, one that is especially strong in women. He highlighted the significance of the "inborn schemata of the infant" or, as he called it later, an innate releasing mechanism. According to Lorenz, the existence of this innate system of behavior toward infants could be deduced from the identification of innate feelings associated with particular objects described as *herzig*, a term that combines the connotations of "sweet," "neat," and "cute" in English. An encounter with such an object releases the instinctive movement of "taking in the arms," as he said he had witnessed in a striking episode. When she was an infant, Lorenz's daughter saw a doll and, in seconds, ran to take it in her arms with a "motherly" expression. The automatic character of the response and the determination shown in a behavior she performed for the first time seemed comparable to the instinctive reactions of animals.[39] Lorenz described his daughter's response to a doll as equivalent to the *Cairina* mother's response to a chick from another species.

Lorenz presented a diagram illustrating the features that release parental behavior. On the left are faces possessing the features that release parental behavior in humans: small heads, round features, big eyes (*herzig* features). They are contrasted on the right with faces that lack those characteristics (see fig. 2.3).[40]

Before World War II, Lorenz had consistently maintained that his views about the role of instincts in social behavior applied to humans as well as to animals. Elaborating on those views, his paper at the 1949 Cambridge symposium claimed that the existence of human instincts was

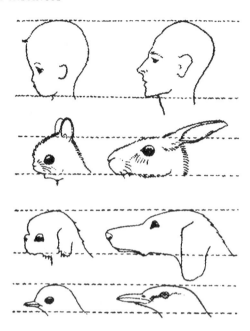

FIGURE 2.3. According to Lorenz, small, round heads with big eyes, bulging craniums, and retreating chins elicit human parental responses; organisms without those features (in the right column) do not. From Konrad Lorenz, "Die angeborenen Formen möglicher Erfahrung," *Zeitschrift für Tierpsychologie* 5 (1943): 276. Courtesy of Konrad Lorenz Archive, Altenberg, Austria.

proved by the existence of human emotions. Here he was following scholars as diverse as Charles Darwin, William James, William McDougall, and John Watson, who had maintained that an emotion always correlates with an instinct, as I discussed in chapter 1. Except for McDougall, Lorenz did not refer to these authors. In his "Companions" paper, Lorenz had written that McDougall had "demonstrated that particular instinctive behaviour patterns are dependent upon specific emotions as subjective correlates." Now Lorenz said he agreed with McDougall's view that "man has just as many 'instincts' as he has qualitatively distinguishable emotions." He then used this correlation as a tool to infer instincts from emotions.[41] This correlation between emotions and instincts explains why we often find Lorenz, as well as other authors, using the terms maternal care and maternal love interchangeably and assuming that if the first is instinctive, so is the second.

Following this line of reasoning, Lorenz argued that parental behavior in humans is innate. Incorporating the material used in his earlier writings, he contended that this was "proven" by the existence of the emotions associated with looking at and interacting with babies. In Lorenz's words:

> It is a distinct and indubitably sensuous pleasure to fondle a nice plump, appetizing human baby.... In this case, the existence of a true innate releasing mechanism in man has been clearly proven.... Also, the objective and subjective reactions activated by the mechanism are clearly distinguishable. A normal man—let alone a woman—will find it exceedingly difficult to leave to its fate even a puppy, after he or she has enjoyed fondling and petting it. A very distinct "mood," a readiness to take care of the object in a specific manner, is brought about with the predictability of an unconditioned response.[42]

As with other instincts, the behavior's emotional quality, the fixity of the response, its universality, and its machinelike character proved the instinctual basis of parental, and especially maternal, behavior.

In 1953, ten years after his long paper on innate forms of possible experience, Lorenz published a condensed but otherwise almost verbatim account of his views about human instincts. In this paper too, Lorenz devoted much attention to parental care, repeated his views about the innate releasing mechanism toward the cute—especially strong in women—and asserted the instinctive nature of mother love as well as monogamous love.[43]

After the war, building on the success of his popular book *King Solomon's Ring*, Lorenz devoted himself to promoting his ideas and reaching large audiences. He published more often in English, had many of his earlier essays and books translated, and welcomed foreign visitors and researchers to his research station. In addition, he traveled extensively in the United States, where in 1954–55 he delivered talks at prestigious universities such as Harvard, Cornell, and Clark University, as well as public lectures in cities like Boston and New York City. He also attended numerous international meetings throughout Europe. In some of those meetings, encouraged by psychiatrists' interest in his work, Lorenz's assertions about instinctive behavior in humans became increasingly bold.

The WHO Meetings: Imprinting from Birds to Infants

Lorenz found an eager audience for his views on social behavior in an international Study Group on the Psychobiological Development of the Child, organized by the World Health Organization (WHO). Chaired by one of the pioneers in studies about children and human emotions, Frank Fremont-Smith of the Josiah Macy Jr. Foundation, this group met in Geneva in 1953, in London in 1954, and in Geneva again in 1955 and 1956. Besides Lorenz, members included the British psychoanalyst and psychiatrist John Bowlby, the American anthropologist Margaret Mead, and the Swiss psychologist Jean Piaget. Guest speakers included biologist Julian Huxley and psychoanalyst Erik Erikson. The meetings focused on developments in ethology and their implications for child psychology.[44]

Bowlby made sure the mother-child relationship would be a focal point of the discussions. In his introduction during the first meeting, he highlighted his interest in ethology and noted that his investigations into the effects of separation from the mother had led him to Lorenz's work. He was interested in finding out if imprinting operated in humans and could help explain infants' negative reaction to separation from their mothers. He also noted that ethological studies of the mother-child relationship should be of interest to psychoanalysts, who had already placed the mother-child dyad at the center of social development. As Bowlby put it, "The phenomenon of imprinting at once struck me as possibly important to my work. Whether it really has anything to do with the effects of separation we shall see. The other thing that fascinated me in [Lorenz's] work was the mother-child relationship of animals. The mother-child relationship is manifestly an example of instinct, in the ethological meaning of the word, and it is also at the centre of psychoanalysis."[45]

As early as 1934, Lorenz had been aware that some psychologists and psychiatrists showed interest in imprinting. He wrote to Stresemann: "I have finally gotten in contact with the psychologists, that is, they now know about me. The psychiatrists also start to be interested in the phenomenon of imprinting. They believe there could be something similar in humans, which I do not think is impossible. But to see it clearly, I would need a series of 6 male and 6 female Kasper Hausers, who will hardly be granted to me!"[46]

Now, speaking before an audience with few experts in biology, Lorenz made daring pronouncements about human behavior. In a memorandum he sent to the WHO regional office a few weeks before the first meeting, Lorenz said he would focus on two processes of interest to students of child development: innate releasing mechanisms (IRMs) and imprinting. He would talk about the existence of IRMs in the human species and deal "with the extreme probability of imprinting in human children." Finally, he would treat the pathological disintegration of IRMs and the pathology of imprinting. Lorenz's presentation on imprinting elaborated on the significance of this phenomenon, which until that point had been observed in only a few species: "Though imprinting has been found in its typical form in birds and insects rather than in mammals, I really do believe it to be fundamentally akin to those very lasting object-fixations of human beings, chiefly because these fixations also seem to be dependent on early childhood impressions and seem also to be largely irreversible. Some psychiatrists and psychoanalysts here I believe share this opinion, at least as a working hypothesis."[47]

During the discussions, Lorenz acknowledged that little could be said with certainty about humans or any mammal: "We don't know a thing about them. . . . Maybe in about five years I can just tell you something about small monkeys, or lemurs, with which we intend to start." Further, he pleaded: "As to experiments, I must ask you not to expect too much knowledge about imprinting in man from ethologists."[48]

Yet Lorenz had already stated his belief in the instinctual nature of human social behavior in no uncertain terms. Here, and in much the same words as in his previous writings, Lorenz asserted that human maternal behavior was a clear instance of innate behavior, as proved not by biology but by social interactions. This is how he put it:

> But now let me proceed to what interests us most, the mother-child relationship. One of the best instances of the I.R.M., except for the snake, is our reaction to the quality of *cute*. . . . Now, let's look at the properties which produce the impression of a thing being *cute*. The head must have a large neurocranium and a considerable recession of the viscerocranium, it must have an eye which is below the middle of the whole profile. Beneath the eye there must be a fat cheek. The extremities must be short and broad. The consistency of the body ought to be that of a half-inflated football, elastic; movements that are rather clumsy elicit the reaction very strongly; and finally the whole thing must be small, and must be the miniature of something.[49]

In short, "cute" equals "baby"—as he had already asserted in his 1943 and 1953 papers.

But how can we know whether parental care toward babies is instinctual? To answer the question, Lorenz appealed to the results of social "experiments":

> Now, in order to see whether many people have got that I.R.M., we ought to do a mass experiment with thousands or millions of experimental persons. Just

this experiment has already been done: it has been done by the doll industry, which, of course, sells the supranormal object best. The exaggeration of key-stimuli can be very nicely shown in the "cupie" [kewpie] doll, and the "Käthe Kruse Puppe" in German, and if you want facts on what I say, then go to Walt Disney's films and see how Walt Disney represents cute animals.[50]

In the supranormal or supernormal object the characteristics that stimulate the release of an innate behavior are exaggerated. For example, Tinbergen had shown that the oystercatcher (a bird) prefers a giant egg, even an artificial one, to a normal egg. In other cases it is not size but some other characteristic of the object that triggers an animal's instinctual reaction.[51]

Lorenz portrayed the appeal of a doll as support for the existence of a female maternal instinct. Yet Lorenz had not conducted research on these industries, nor did he present data on these issues from other sources. In addition, he did not consider any alternative explanations for the behaviors he was discussing. For example, he did not examine the possible role of environmental influences on maternal behavior or attitudes. By this point several researchers, including the American psychologist Leta Hollingworth and the sociologist Ruth Reed, had done extensive research showing that society's emphasis on women being nurturing, their roles as caretakers, and social expectations that they should be "maternal" all influenced their interest in babies.[52]

Lorenz, however, asserted not only that maternal behavior is instinctive but also that the value societies place on such behavior is innate:

> We must keep in mind that mother-love is not more necessary to the survival of the species than the drive to copulation. Why, then, are those drives to copulation "brutish" and why is "maternal love" sublime? This is simply our emotional valuation of instinctive behaviour in man—and it is largely dependent on supply and demand. I am convinced that we have something very deep, innate, in our behaviour, which tends to devalue sex and eating and to value very highly mother-love, social behaviour, defence of the family, and so on.[53]

Anthropologist Margaret Mead, who became a close friend of Lorenz but was also one of the few people in these meetings who sometimes criticized his views, pointed out that no such universal valuation exists, since one could find "societies which put a high value on sex and eating, and a low value on maternity."[54] Lorenz did not respond to her objection.

When he turned to parental behavior during the third meeting, held in 1955, Lorenz maintained that there was only a quantitative difference between men's and women's reactions to babies. Only cultural mores prevented "giving utterance to these, certainly instinctive, urges" in males. However, he also proposed that the urge to develop different gender roles is innate:

> Well, I had better come out and be honest about what I am aiming at. I do believe that there is a certain unlearned element—something like an IRM—which makes the little boy actually seek for somebody to take over the father role. Sylvia Klimpfinger has evidence for that in a hospital—a hostel—where all the children are reared by the female staff alone, and all these children—the boys more significantly than the girls—go for the gardener who is the only male accessible to them. This led me to suspect that there might be an unlearned preference for what to imitate—boys to imitate Pa and for girls to imitate Ma.[55]

In sum, after asserting that biologists knew nothing about the biological basis of behavior in mammals, let alone humans, Lorenz argued that parental and sexual roles are innate in humans, as are their ethical valuations of those roles. According to him, humans possess an internal releasing mechanism for parental behavior toward babies, as shown by a universal tendency to consider baby features cute. In addition, people instinctively give a high moral value to mother love. Finally, there is an innate preference for boys to imitate their fathers and for girls to imitate their mothers. The combination of those points amounted to an argument for the instinctual basis of traditional gender roles and, specifically, gendered parental roles.

Since Lorenz did not devote specific books or articles exclusively to this topic, one might be tempted to conclude that his views about maternal care were not central to his career. On the contrary, I believe they played an important part in his success, especially in the United States. His role as an expert on maternal care transcended the confines of academic meetings and was key to his prominent visibility in different disciplines and among the general public. As historian Gregg Mitman has shown, Lorenz's and also Tinbergen's focus on the family life of animals was the center of their films and media work.[56] In the United States, many of Lorenz's public appearances emphasized his expertise on "motherhood" in ducks and humans. Magazines and television programs also focused on his ability to substitute for a bird's mother. His "mothering" abilities made him appealing to large audiences. When *Life* magazine run a story about

FIGURE 2.4. Lorenz as "An Adopted Mother Goose." The caption of the picture in *Life Magazine* reads: "Standing chest-deep in water on his Westphalia preserve, Dr. Lorenz assembles his family of goslings by making noises like mother goose." *Life Magazine* 39 (July/August 1955): 73. Photo by Thomas D. McAvoy. TIME & LIFE Images. Courtesy of Getty Images.

Lorenz in 1955, for instance, the title presented him in his most popular role, as "An Adopted Mother Goose" (see fig. 2.4).[57]

As an adopted mother, Lorenz enjoyed a privileged position from which to observe the social life of animals. In his best-selling *King Solomon's Ring*, Lorenz detailed his role as a devoted "foster mother" of jackdaws, ducklings, and goslings. His success in this role conferred authority on his views concerning maternal care and even infant development. Thanks to his ability to "talk to the beasts, the birds, and the fish," Lorenz, like King Solomon, could then share the animals' wisdom about mother love. His lessons seemed to come straight from nature, right from the goose's mouth.

At the same time, Lorenz used his intimate knowledge of animal family life to extrapolate from birds to humans. This extrapolation was especially successful in the case of his work with geese. As Klaus Taschwer has put it, geese were Lorenz's emblematic animal.[58] Lorenz's best-known images show him swimming with or being followed by geese. Observers could readily see him as part of their family life. In a fascinating examination of his changing representations of his "goose child Martina," Tania Munz has shown how Lorenz used geese, and specifically Martina, to instantiate different—and sometimes contradictory—views over the span of his career.[59] Lorenz literally attributed to Martina those characteristics and pathologies that he believed were typical of animal and human behavior. In this sense, when Lorenz talked about animals, he was often talking about humans as well.

Lorenz used several strategies to achieve this identification of humans with animals, especially geese. First, he established the equivalency of humans and animals at the emotional level. He was fond of citing his mentor Heinroth's words: "I regard animals as very emotional people with very little intelligence!"[60] He also often emphasized that greylag geese exhibited social and especially family behaviors similar to those of humans. The strong pair bond leading to monogamy, the healthy family, the coy female that did not engage in indiscriminate mating: these were all features of the healthy family that Lorenz saw as the fundamental pillar of a stable society, in geese and in humans.

In addition, Lorenz used numerous rhetorical strategies to promote the identification of humans with his animals. He gave the animals he raised human names, like Martina. Sometimes he told the life stories of individual animals, like the greylag goose Martina and the jackdaw Tshock. These biographies were built around the moments of significance in a human life—birth, attachment to the mother, falling in love—and included all the dramatic elements of a human life: jealousy, overcoming obstacles, faithfulness, courage, and so forth. The animals played humanlike roles: parent, groom, bride, sibling, or child. They lived humanlike lives, and a human playing the role of an animal (Lorenz) narrated their adventures. For Lorenz, the human-animal boundary was highly permeable.

All those elements helped readers empathize with the geese, and they also provided evidence for Lorenz's contention that, emotionally, humans are like animals. If we are like animals, for him that also meant that much of our emotional equipment is inborn and much of our social behavior is determined by instincts. As I said earlier, Lorenz had already stated in his early writings that social behavior in higher animals is to a large extent determined by instincts. This bold assertion was made explicit and more appealing by illustrating how those instincts work in the lives of animals.

Lost in translation between the scientist and the foster mother, between the geese and the humans, was an important aspect of imprinting that remained hidden in many subsequent discussions. Birds do not imprint on an individual, but on the species through an individual. Lorenz's mentor Heinroth, who first highlighted the phenomenon, always talked about how geese imprint on humans and then see humans as their parents. Lorenz also wrote that imprinting provides the image of the species to the animal imprinted. However, he always emphasized his personal relationship with a bird imprinted specifically on himself. When Lorenz wrote about those birds that followed him, he always noted that they considered *him* their mother. For example, as Munz has noted, he presented his most famous goose, Martina, as "my goose child."[61]

Lorenz's emphasis on his particular relationship with a bird imprinted on him helped to make imprinting in animals parallel to the human mother-infant dyad. In this sense, it facilitated the extrapolation of imprinting from animals to humans. To a great extent, I believe, this displacement from the species to the individual made it easier for psychoanalysts like Bowlby to see imprinting as analogous to the relationship of human children to their mothers.

Conclusion

Konrad Lorenz and Niko Tinbergen presented ethology as a new science that would focus on the biological study of instincts. They argued that human social behavior, like the social behavior of other animals, is guided by instincts. After World War II, at a time of heightened scientific and public interest in the development of emotions and the mother-child relationship, Lorenz's studies of imprinting and animal social life received much publicity. For many psychoanalysts and psychologists, including John Bowlby, several psychopathologies and sociopathologies could be traced back to the disruption of the mother-infant bond. To them, Lorenz's work on imprinting seemed relevant to understanding the nature of early relationships.

Here I have shown how Lorenz extrapolated his views about animals to humans and, specifically, endorsed the relevance of imprinting for understanding child development. Encouraged by Bowlby and other child researchers' eager reception of his views, Lorenz increasingly presented himself as an expert on the mother-child dyad. At least in the United States, Lorenz's success was closely tied to his role as "an adopted mother goose."

During the 1950s many child analysts, in turn, would use Lorenz's ideas to support their own views. One of the most prominent was Bowlby, who—with Lorenz's endorsement—gradually turned to animal research to back his view that the mother is the child's psychic organizer and in 1958 proposed a synthesis of ethology and psychoanalysis to explain children's attachment to their mothers.

CHAPTER THREE

Bowlby's Ethological Theory of Attachment Behavior

The Nature and Nurture of Love for the Mother

There is a growing realization that the way a mother feels about her child is the most significant feature of all in a child's development and in his attitude to those around him. —John Bowlby, "The Rediscovery of the Family" (1954)

Introduction

After his 1951 WHO report, John Bowlby continued developing his views about the instinctual nature of the mother-child relationship and the disastrous effects that maternal separation and deprivation have on children. Since Bowlby was British and spent his career in England, most historical research has focused on his influence within the English context. In addition, a large number of important works on the history of child rearing, adoption, the family, and changing conceptions of motherhood in the United States have briefly noted the impact of Bowlby's work, identifying him as the most important figure in postwar debates about the effects of maternal deprivation. Yet there is no study of Bowlby's influence in the United States.[1]

Here I examine the development of Bowlby's views and their scientific and social reception in the United States during the 1950s, a pivotal period in the evolution of his views and in debates about the social implications of his work. Bowlby's assertion that mother love is a biological need for children influenced discussions about whether mothers should work outside the home and supported a gendered division of parental care. Bowlby's influence, I propose, was heightened because the social interest in his views about the effects of maternal deprivation came at a crucial juncture in debates about women's role in modern society, and also because of the emotional effect his position had on mothers.

Not only was Bowlby important in the United States, but the United States was crucial for Bowlby. American agencies funded some of his research at the Tavistock Clinic. Bowlby also spent the 1957–58 academic year at the Center for Behavioral Studies at Stanford, a period seminal in the development of his ethological theory of attachment behavior. In addition, many researchers whose work Bowlby relied on were American, as were many who criticized his work.

Some American psychologists criticized the lack of convincing evidence that maternal deprivation had severe and lasting consequences for children's emotional development. These criticisms prompted Bowlby to rely heavily on animal behavior studies to support his views. As we saw in chapter 2, Lorenz encouraged his expectations that ethology would help him flesh out a biological account of the mother-child dyad. Bowlby established a naturalistic framework for the mother-infant dyad by combining specific ideas from psychoanalysis and from ethology. In his classic 1958 paper "The Nature of the Child's Tie to His Mother," Bowlby presented his ethological theory of attachment behavior. He argued that attachment to the mother is an innate biological need and consequently that separation from the mother or lack of mother love has catastrophic consequences for a child's development.

By claiming biology proved that a mother's heart determines the mind of her child, Bowlby's argument placed an unusually strong emotional and moral demand on mothers. As historians such as Ruth Bloch and Jan Lewis have shown, the power of mother love in shaping a child's mind is "the kernel of the emotionology of motherhood" in the United States. But this power was now shored up with the authority of biological science, since Bowlby presented mother love as a child's biological need and claimed the support of ethology for this view.[2]

From Natural Description to Social Prescription: Infants' Needs and the Tragedy of Working Mothers

When Bowlby's 1951 WHO report and its more popular 1953 version appeared, his main thesis about the importance of mother love fit well with the consensus developing within the American psychoanalytic community, as we saw in chapter 1. In fact Bowlby had already drawn on the studies of some American researchers, including David Levy and René Spitz, to support his views in the WHO report. However, Bowlby's study was not just one more work on maternal deprivation, for three reasons.

First, in his report Bowlby actively constructed a consensus by strategically emphasizing the common points among the researchers working on children. He relied on studies that covered a range of related but distinct issues: infants separated from their families owing to hospitalization of the mother or the infant; separation from the mother for short periods; permanent separation from mothers or the whole family; unsatisfactory maternal care; and faulty emotional attitude of the mother toward the child. But he included all those studies under the umbrella of "maternal care and love," obscuring the dissimilarities among the children and situations studied. In addition, he boosted this consensus by appealing to a standard epistemological tenet in science. He noted that one particular study on a topic cannot provide convincing proof of a conclusion, but the convergence of several studies done independently adds evidential support to individual conclusions. Thus his views were proved by the convergence of similar results "from many sources." This convergence, he often repeated, left "no doubt that the main proposition is true."[3]

The second reason is that Bowlby's WHO report, as a document backed by a respected international organization, gave his views a visibility and respectability that none of the previous studies enjoyed independently. On publication, it became the authoritative document of the consensus within a large field of research. The ideas of René Spitz, David Levy, Margaret Ribble, and Therese Benedek were well known among child analysts, but with Bowlby's report they gained greater visibility beyond psychoanalytic and psychiatric circles. Although Bowlby was not the only scientist moving toward a deterministic view of mother love, his views epitomize its strongest instantiation, and he became its most visible advocate.

The third reason is that Bowlby, more than any of the other researchers in the field, was active in drawing practical implications from his research. Bowlby had concluded his report with a clear prescription for mothers:

> The provision of constant attention night and day, seven days a week and 365 days in the year, is possible only for a woman who derives profound satisfaction from seeing her child grow from babyhood, through the many phases of childhood, to become an independent man or woman, and knows that it is her care which has made this possible.[4]

Furthermore, Bowlby continued, for the mother to provide constant devotion she needs a support group—the family.

For him the family was a "natural unit," essential for children's healthy
development: "It is for these reasons that the mother-love which a young
child needs is so easily provided within the family, and is so very, very difficult to provide outside it."[5] Bowlby referred to the family as the "natural
home group," and he compiled a list of the reasons that the natural home
group might fail to care properly for the child: illegitimacy, chronic illness,
economic conditions, war, famine, death of a parent, desertion, imprison-
ment, divorce, and full-time employment of the mother. "Any family suffering from one or more of these conditions must be regarded as a possible source of deprived children," he argued.[6]

Could "full-time employment of the mother" really be equivalent to
famine, war, or death of a parent? This was a simple corollary of Bowlby's
position. If, as he believed, the constant presence of a loving mother is
necessary for the child's mental health, her absence would be catastrophic.

The catastrophe was social as well as personal, according to his analysis. Lack of mother love led not only to psychopathology but also to
sociopathology. "The proper care of children," Bowlby continued, is "essential for the mental and social welfare of a community." And "when
their care is neglected, as happens in every country of the Western world
to-day, they grow up to reproduce themselves. Deprived children, whether
in their own homes or out of them, are the source of social infection as
real and serious as are carriers of diphtheria and typhoid."[7] Thus, in ful-
filling their natural role, women were doing the right thing toward their
children, their families, and their communities. Mental health, achieved
through emotional well-being, was as important for the social body as
physical hygiene was for the organism. The health of the social body was
thus in the hands, or rather the hearts, of mothers.

Through this line of argument, Bowlby revealed his concern not only
about children, but also about group welfare, especially the maintenance
of a distribution of parental roles that he believed would ensure the traditional social order and the very continuity of the species. In scholarly
writings and conferences, Bowlby presented this position as conclusively
proved by scientific research. From his standpoint, the implications were
clear. A mother needed to stay home, for the sake of her child and society.

Through the WHO report and its coverage by the international press,
Bowlby's ideas spread rapidly among scientific and public audiences.
When Bowlby titled his piece in *Home Companion* "Mother Is the Whole

World," the message reverberated across the globe. The *Johannesburg Star* in South Africa reported Bowlby as saying that "when deprived of a mother's care a child's development is almost always retarded physically, intellectually and socially." Another South African newspaper, *Cape Argus,* reported that "Social Behaviour Depends on Mother Love." "The Importance of a Mother's Love" was also covered in the *East African Standard* from Nairobi. The French reported that between the attitude of the mother and the future behavior of her children, there was "une relation de cause à effet quasi mathématique" (an almost mathematical relationship of cause and effect). The Italians put it more poetically: "Solo lé mani di una Madre possono plasmare il destino" (only the hands of a mother can shape destiny).[8]

For American women, Bowlby's message about the crucial role of mother love in children's development and the importance of the "natural home group" arrived at a time of major changes in women's roles and widespread social concern about the rising number of women working outside the home. After World War II, the patriarchal family gained support from numerous measures. As historian Nancy Cott has shown, the social benefits of the 1944 GI Bill of Rights, which assumed responsibility for veterans' economic well-being, helped to enhance "men's roles as husband-heads of households, as property owners, as job-holders and providers," since women were only about 2 percent of all military personnel. In addition, after the war the government showed little interest in continuing the limited support for child-care centers that had been provided during the war, when one and a half million mothers of small children entered the workforce to support the war effort.[9] Cold War propaganda also reinforced traditional gender roles within the American family, as historian Elaine Tyler May has documented.[10] When mounting domestic conflicts and international tensions grew, public and government rhetoric promoted "togetherness" as the key to security. All these measures encouraged the traditional separation of gender roles, with women at home and men in the workforce. A baby boom and sprawling suburbs attest to the impact of these measures of containment, even if, as the Cold War went on, "the rush to marry and buy homes, the reinscription of traditional gender roles, and the overinsistence on the pleasures of family life" revealed less "signs of self-satisfaction than defenses against uncertainty," as historian Gaile McGregor has argued.[11]

In this anxiety-filled environment, two social indicators raised concern about the malfunctioning of families and resulting harm to children. First, the numbers showed that postwar men and women were not adjusting to each other easily. Whereas in 1940 one marriage in six had ended in divorce, by 1946 one in four did. A million GIs were divorced by 1950.[12] The rising divorce rates fueled concerns about the social impact on children and adolescents. Second, the Children's Bureau record of Juvenile Court cases and the FBI compendium of police arrests pointed to a steep rise in juvenile delinquency during World War II, followed by a sharp decline and then another rise during the 1950s. Gallup polls and popular articles revealed increasing alarm about juvenile delinquency, the rise of gangs, and the decline of parental guidance. In 1953 the US Senate began extensive investigations about juvenile delinquency that lasted over a decade. These events helped sensationalize the issue and turn it into a national crisis. While politicians debated and called for data, models, and experts, the young drove the issue out into the open. Marlon Brando riding a motorcycle in *The Wild One* and James Dean cruising in a sports car in *Rebel without a Cause* became emblematic of reckless American youth. Their unorthodox ways raised fears of impending social and moral decay propelled by emotionally unstable adolescents.[13] In turn, heightened worries about juvenile delinquency drew further attention to dysfunctional families.

Growing fears about higher divorce rates and juvenile delinquency encouraged debate about the role of women in the new social order. The extent of this mid-century debate can be appreciated by looking at the rise of studies about women in the mid-fifties. Let's pick 1953, the year of Bowlby's best seller on maternal care and the initial Senate hearings on juvenile delinquency. That year alone saw the publication of the following important works: Alfred Kinsey, *Sexual Behavior in the Human Female*; Mirra Komarovsky, *Women in the Modern World*; Ashley Montagu, *The Natural Superiority of Women*; and the American translation of Simone de Beauvoir, *The Second Sex*. This selection gives us a sense of the expanding body of work on women's issues.[14] These academic treatises did not, however, provide a uniform answer to questions about women's nature and their functions in modern society.

Some contemporary studies and historical works have underscored the conflicting nature of the messages sent to women during the 1950s. As I noted in chapter 1, Komarovsky's earlier sociological research showing that contradictory social messages to be autonomous, independent, and mature, on the one hand, and dependent, subordinate, and childlike, on the other, created deep internal conflicts for many women. In her analysis of women's popular magazines from 1946 to 1958, historian Joanne Meyerowitz has documented the conflicting messages women received during this period as well. Some of these glorified domesticity, while others advocated individual striving and public service.[15] Increasingly, historical research on the 1950s has questioned the common images of domestic bliss and complacency about traditional gender roles. Scholarly studies on the nature and roles of women presented conflicting views that exposed the tensions bubbling beneath the superficial image of happy suburban domesticity. After all, if women were "naturally superior," as Montagu claimed, why were they treated as "the second sex," as Beauvoir argued?

As a short analysis of the reactions to *The Second Sex* reveals, at the core of these concerns about women's changing roles were deep anxieties about motherhood. Written by one of France's most important postwar philosophers, this tract offered an encyclopedic review of historical events and ideas that had led to the construction of woman as "the Other," the opposite of the male and masculine. The other is not only different, but inferior. As in any work of grand scope, many of Beauvoir's points could be debated. Yet almost all American reviewers, male and female, housewives and scholars, scientists and humanists, focused on her assertion that "no maternal 'instinct' exists" and what they saw as her denigration of motherhood.[16]

The concern about this issue reflected the widespread uneasiness created by the greatest increase in American history in mothers of young children going to work outside the home. By the mid-1950s a "silent revolution" had occurred, noted sociologists Alva Myrdal and Viola Klein. More and more married women of different social classes were entering the labor force. At first they sought paid work out of economic necessity and to support the war effort. But now, explained Myrdal and Klein, "the economic motive can no longer be separated from the ideological one; nor can the voluntary element be distinguished from the compulsory one." Between 1940 and 1960, the number of married women with paid jobs doubled, and working mothers increased by 400 percent. Over four million married women took jobs, accounting for 60 percent of all new workers. In 1940 working wives were mainly working class; by the end of the 1950s, many were educated and middle class.[17]

Because more mothers were joining the workforce, the need for child care had again nearly become a crisis even before the Korean War began in 1950, according to historian Sonya Michel. In 1953 the US Department of Labor published *Employed Mothers and Child Care*, a report on "a subject of vital national interest at a time when married women constitute the largest labor reserve in the country, and therefore may be expected to continue entering the labor force in ever-increasing numbers, and when 5¼ million mothers already are employed." Of those, two million were reported to have children less than six years old.[18]

In this context, research about the effects of maternal separation and deprivation attracted great interest. In her study of the child guidance movement, Kathleen Jones has identified a prewar shift that emphasized the psychology of the individual rather than the social networks and circumstances of "the troublesome child." After the war, congruent with the postwar American romance with psychology documented by historian Ellen Herman and the more general interest in social engineering examined by historians such as Mark Solovey, social and political decisions about day care and child rearing increasingly became framed as empirical questions about the emotional needs of children.[19]

Bowlby was thus addressing one of the major concerns of the postwar period in the United States. Backed by a prominent world organization, his views became a point of reference in discussions about the family, personality formation, and parental roles. Many social scientists in the United States had met Bowlby personally when he visited the country to elaborate on the WHO report. This report and its 1953 popular version were also widely reviewed in American journals.[20] In addition, one of Bowlby's closest collaborators, James Robertson, presented their work to American audiences through an influential film.

Robertson had worked at a number of odd jobs at the Hampstead Nurseries before moving to the Tavistock Child Development Research Unit in London to work with Bowlby in 1948. Supported by Bowlby and Anna Freud, he then applied to the Institute of Psychoanalysis for training and became an associate member of the British Psychoanalytical Society in 1953.[21] Under Bowlby's supervision, Robertson had started a project to observe the reactions of children separated from their families by hospitalization. Impressed by the effects on children and aware that his views might encounter resistance in the medical community, Robertson decided to make a film. In 1953, with support from the WHO, he took *A Two-Year-Old Goes to Hospital: A Scientific Film* on a six-week tour of the United States. The Children's Bureau in Washington, DC, worked out the itinerary, which included Yale and the psychoanalytical societies in Boston and New York.[22] The film was silent but was showed with a guide that listed the work being carried out in Bowlby's unit. At this point, Bowlby and Robertson were also writing a book about the effects of separation together with American psychologist Mary Ainsworth. Thus, Robertson's presentation of his film on separation in hospitals also helped disseminate Bowlby's views in the United States. In a 1954 letter to Anna Freud, Robertson told her that the film's best reception had been in America.[23]

Other indications of American interest in Bowlby's work come from his funding sources and connections. The Josiah Macy Jr. Foundation and the Ford Foundation generously supported the work that Bowlby's team carried out at the Tavistock Institute. Bowlby also held a fellowship at the Center for Advanced Study in the Behavioral Sciences in Stanford, California, in 1957–58. During his stay, he received numerous invitations to speak in different forums and visited several research centers in the country. Actively engaged with a circle of researchers that included many American psychologists, animal researchers, psychoanalysts, and workers in child development, Bowlby exerted influence well beyond England and the psychoanalytic establishment.

Although in the early and mid-1950s Bowlby's views were still pretty similar to those of other researchers, his status as the scientist behind the WHO report and his willingness to extract social prescriptions from his work made him a central reference in debates about women's work, maternal care, and children's emotions. His work was discussed widely, in policy conferences and the public media.

The Midcentury White House Conference on Children and Youth brought together Benjamin Spock, David Levy, Erik Erikson, Ashley Montagu, and other American psychologists, psychiatrists, and social scientists to address how children develop a healthy personality. The fifth in a series of decennial conferences started in 1909, it focused on "how to rear an emotionally healthy generation" and called attention to the children's "feelings." Earlier conferences had focused on the economic and social aspects of the problems American children encountered. Now, although the report from the meeting recognized that "emotional ill health may have economic, sociological, physical, psychological, and spiritual causes," it nevertheless underscored that "some of the chief ills of the present day are psychological."[24]

The conference also aimed to extract the policy implications of research on children's needs. One explicit goal was to find out the latest in research so that someone like Spock could then put it into a form understandable to the general public. Despite noting the tentative character of this knowledge and recognizing that there were several competing theories, the report claimed it was "well established that loving care is essential for the well-being of children." Both the discussion during the meetings and the final report focused largely on Erikson's views about child development, especially the importance of trust as a basic component of a healthy personality. Erikson, who had achieved national prominence with his *Childhood and Society*, emphasized that maternal love enables the child to develop a sense of self.[25] To support the view that mother love is determinant during infancy, the report referred to Bowlby's WHO report. Later, in the section "Effects of Deprivation of Maternal Care," it also appealed to Bowlby's WHO publication and quoted approvingly and at length his views about the detrimental effects of lack of mother love in infancy.[26]

Important scientists from other fields, including sociology and anthropology, appealed to Bowlby's work to defend the thesis that the child has an innate need for mother love. In "The Power of Creative Love," Harvard sociologist Pitirim A. Sorokin and his student Robert C. Hanson argued that motherly love is vital for babies. Even anthropologist Ashley Montagu, now best remembered for his opposition to biological explanations of behavior, argued that when researchers studied individuals incapable of showing love, it was "invariably found that something was lacking in their mother's relationship to them." Though Montagu had done no research on this topic, he recommended Bowlby's "admirable analysis" in the WHO report. Montagu also reported that children without love die, this time noting the investigations of David Levy. As an anthropologist, Montagu pointed to a variety of forms of love in different cultures, but he then claimed that they all would be "traceable to the need for the kind of love which is biologically determined, predetermined, to exist between mother and infant." The implications for mothers were serious: "To the extent to which women succeed or do not succeed in adequately loving their children, the boys and girls become inadequately loving men and women."[27]

The influence of Bowlby's views is also exemplified by Myrdal and Klein's treatment of the topic in *Women's Two Roles*. In the seemingly obligatory chapter on the "Effects of Mother's Employment on the Mental Health of Children," they discussed Bowlby's work and argued that "it would be scientifically inadmissible to apply conclusions drawn from cases of deprivation caused by emergency situations, such as death, abandonment or cruelty of the mother, or the separation through illness of mother or child, to cases where the mother is absent at regular intervals for a number of hours yet returns to the child each day and provides it with a home." They also denied the scientific validity of a lot of research on maternal deprivation and questioned its application to normal families with working mothers. Nevertheless, they concluded: "All we can do at present is to stress the undeniable fact that maternal love is a decisive element in any equation concerning young children." And later: "We therefore support the view that mothers should, as far as possible, take care of their own children during the first years of their lives."[28]

But could mothers' working outside their homes really endanger the health of their infants and even the whole nation? According to Bowlby, children's lack of mother love in their early years results in a number of social problems. He claimed that delinquency, among other things, could stem from a lack of maternal care. Was the rise of delinquency in America perhaps evidence that mothers were slacking? He included divorce as a problematic disruption of the natural home unit. And wasn't divorce becoming more prevalent in American society? Even more worrisome, Bowlby had presented "full-time employment of the mother" as equivalent to famine, war, or death of a parent.

Given that more and more women were entering the workforce, the question acquired social urgency: "Should a woman with children take a job?" asked the English *News Chronicle*. "The mother who stays at home gives her children a better chance," answered Bowlby.[29]

In the *New York Times*, Sloan Wilson, author of the best-selling novel *The Man in the Gray Flannel Suit* (later made into a successful movie starring Gregory Peck), contributed to the public debate with his declaration that married women with children should not have business careers. Instead, they should assume "executive wifehood," helping their husbands' careers. Bernice Fitz-Gibbon, the 1956 Woman of the Year in Business, mother of two and grandmother of three, responded that a career could make women better wives and mothers. So she encouraged women to take the "gay" rather than the gray flannel suit. In the ensuing debate, Wilson cited psychiatrists' call for a "loving mother who has plenty of time to give her sons and daughters." David R. Mace, a professor of human relations, posed a simple question to Fitz-Gibbon: Had she ever heard of "Dr. John Bowlby, a psychiatrist of international reputation, whose impressive and well-documented report to the world's mental-health experts named maternal deprivation as a major cause of serious personality disorders?"[30]

The *Ladies' Home Journal* continued the discussion with a forum titled "Should Mothers of Young Children Work?" Besides Bowlby himself, the panelists in the forum included US secretary of labor James P. Mitchell, sociologist Mirra Komarovsky, Dr. Lynn White, a Jungian lay analyst, a mother and grandmother named Mrs. Florida Scott-Maxwell, Mrs. Roy Davis, a nurse and mother of four, and others. The article reported that while traditionally women had worked only out of economic necessity, at present they were going to work "by choice." But was this the right choice?

Although Secretary Mitchell argued that American women needed to be part of the workforce to maintain their current standard of living and to contribute to the national defense, he also claimed that "no nation should ever forget that the very primary, fundamental basis of a free society is the family structure—the home—and the most vital job is there." Should mothers, then, be denied the choice to work? In the midst of a Cold War in which the United States held up individual freedom as the basis for its superiority, Mitchell was not prepared to deny American women this freedom: "I think it is very right that we in this country have freedom of choice, unlike the Communist world, where there is no such thing." Nevertheless, he hoped that women workers would not be mothers, since the mother's place was "in the home." Mitchell thus defended the superiority of the American model by the somewhat ironic position that American women were free to do the wrong thing.

The women in the forum presented different viewpoints. Some who worked and had sitters reported that their children were doing fine. Others who stayed home felt that working women looked down upon them. Some said working was not always a matter of choice, for their families could not live well if they did not work. It was not a simple decision between the apron and the gay or gray flannel suit.

As for Bowlby, he repeated the advice he believed followed from his scientific work: "To deprive a small child of his mother's companionship is as bad as depriving him of vitamins." Seemingly aware of the boredom of suburban mothers, he recognized that most women "would like a more varied life than is available in the modern suburb," where "we have made it almost impossible for them to take care of their children happily and to combine this with some sort of career or job." His solution, though, was not to change social practices but to change social values, so that the home was given its proper place. To do that, "we must first ascertain who it is that holds the values we oppose. Personally, and this is pure prejudice, I think it is career women who look down on women who stay home." He further argued that group care was not mothering. Neither should children under three go to day care—though after that age, he realized, "part-time day care has its uses." Still, he insisted that children deprived of mothering would grow up to hate and mistrust, leading to a life of truancy and promiscuity.[31]

In turn, work by Bowlby and others on the effects of maternal deprivation helped support a division of parental roles, and consequently gender roles. For example, Ashley Montagu, a staunch supporter of Bowlby's work, contended in "The Triumph and Tragedy of the American Woman" that American women mistakenly believed equal rights implied equality of function. According to him, things were better in Europe, where a woman's life centered "upon the happiness of her husband and children, and this is likely to be satisfying to everyone concerned." Montagu, who had earlier defended the "natural superiority of women," now presented his interpretation of recent research on mother love to a general audience: "I put it down as an axiom that no woman with a husband and small children can hold a full-time job and be a good homemaker at one and the same time." For everybody's sake, Montagu hoped American women would realize that "being a good wife, a good mother, in short, a good homemaker, is the most important of all the occupations in the world."[32] Thus, some social scientists used children's alleged need for constant mother love to justify separate spheres and to reject what Montagu called "the equality of functions."

It is difficult to gauge how far studies of maternal deprivation and the WHO report influenced public opinion. According to historian Sonya Michel, the psychological discourse on maternal deprivation, including Bowlby's, represented the most vehement and explicit opposition both to maternal employment and to child care.[33] In the late 1950s, a National Manpower Council report showed that most Americans agreed that women with small children should not be working.[34] At the very least, it seems safe to assume that the scientific work helped shore up longstanding beliefs about infants' need for their mothers.

Yet during this period, several child psychologists working on maternal deprivation also began to question the validity of many of the studies Bowlby had relied on in his WHO report.

Challenging the Studies on Maternal Deprivation

Bowlby presented the notion of the mother as the psychic organizer as an empirical fact, though he recognized that research on this matter was inconclusive. In the WHO report, he rejected the studies that did not support his position, saying that only "three" studies presented evidence challenging his conclusions and that they lacked "high scientific quality."[35] Yet Bowlby also admitted that there were "still far too few systematic studies and statistical comparisons in which proper control groups have been used" that supported his views about mother love. He recognized that "relatively few studies taken by themselves are more than suggestive" and, furthermore, that there were different interpretations of the data. For example, psychologist William Goldfarb had asserted that deprived children "craved affection," whereas Bowlby described them as "affectionless." But the difference, according to him, was "probably more apparent than real." He also reported that some deprived children stole, but others didn't, as well as that two-thirds of the children deprived of mother love turned out to be "socially capable." This statement seems odd. If maternal love is as essential as vitamins and proteins, how could one explain these findings? But Bowlby rejected those results, saying that experts had not examined those children carefully. Perhaps "psychological troubles not leading to social incompetence were not recorded."[36] He did not specify what other "psychological troubles" he had in mind.

In spite of Bowlby's dismissal, there were significantly diverse interpretations of the data, as well as different views about the social implications of the results. Harvard psychiatrist Abraham Myerson implored scientists to "quit blaming mom," stressing that there was no scientific proof for considering mothers to be the cause of their children's neurosis. Based on an empirical study of 162 "farm children of old American stock," University of Wisconsin sociologist William H. Sewell found no correlation between infant training and personality development.[37] Furthermore, child psychologists were questioning the validity of Margaret Ribble's and René Spitz's work. Samuel Pinneau of the University of California presented the most comprehensive critical analysis.

Pinneau dealt first with the physiological evidence Ribble had provided to support her thesis that children whose mothers were not attuned and responsive to their emotional needs would develop gastrointestinal disturbances, tension, respiratory problems, anxiety, and disorganized neurological function. After examining dozens of studies, some of which failed to confirm Ribble's claims while others refuted them, Pinneau agreed with the conclusion of Harold Orlansky, a Yale anthropologist who had earlier made an extensive critical review of child studies. Pinneau adopted Orlansky's strongly worded assessment of Ribble: "It is unfortunate that such an influential writer has not attempted to draw a line between her empirical findings and her personal opinions."[38]

In his 1953 presidential address to the New York State Psychological Association, child psychologist L. Joseph Stone recognized the effectiveness of Pinneau's critique of Ribble by comparing its power to "a kind of hydrogen bomb." Owing to its awesome "destructive" impact, "not a paragraph is left standing for miles around." But although Stone agreed with the dismissal of Ribble's work, he still found the studies by Spitz and Bowlby convincing.[39]

But then Pinneau published a devastating critique of Spitz's work. Spitz had argued that infants separated from their mothers for over six weeks develop psychogenic disorders and literally wither away and die. Pinneau pointed out that it was difficult to evaluate Spitz's claims because he had not identified the specific sites of his studies—he only mentioned a nursery in a penal institution for delinquent girls and a foundling home. More troubling still, by comparing data Spitz offered in different reports about the same set of studies, Pinneau exposed great inconsistencies and shortcomings: Spitz had not specified the composition and training of the research staff, and he had presented contradictory data on the number of children studied and the parents' characteristics.[40]

In addition, Spitz had not determined the health of infants before or during his research, although he noted a lethal measles epidemic during his study of hospitalism. Owing to the loss of children through adoption, there was also a selective sampling bias in the foundling home. Furthermore, from the evidence presented, it was clear that the groups being compared differed substantially in economic background, constitution, and hereditary makeup. Pinneau pointed out that Spitz was inconsistent in presenting data about when the children were observed, and that he provided only cross-sectional but not longitudinal data. Pinneau also attacked the validity of the developmental scale Spitz used to obtain developmental quotients. After his thorough critique, Pinneau concluded: "The results of Spitz's studies cannot be accepted as scientific evidence supporting the hypothesis that institutional infants develop psychological disorders as a result of being separated from their mothers."[41]

As the evidence mounted against the view that maternal separation has devastating consequences on children, even researchers in Bowlby's own group reconsidered its effects. A 1956 paper by several members of his Tavistock group titled "The Effects of Mother-Child Separation: A Follow-up Study" noted that "some of the workers who first drew attention to the dangers of maternal deprivation resulting from separation have tended on occasion to overstate their case. In particular, statements implying that children who experience institutionalization and similar

forms of severe privation and deprivation in early life *commonly* develop
psychopathic or affectionless characters are incorrect."[42] This statement
caused a stir, since it contradicted Bowlby's position in earlier papers and
in the WHO study.

But Bowlby did not retreat. Almost two years later, responding to letters from perplexed readers who wondered whether he had changed his
views, Bowlby published an explanatory note. He said he wanted to "discourage anyone from supposing that I have changed my position in any
material way."[43] Instead, like child analysts Benedek, Levy, Spitz, and Ribble, who emphasized the natural foundation of the mother-child dyad,
Bowlby turned to investigate its biological foundation.

Uniting Psychoanalysis and Ethology:
The Nature of the Child's Tie to the Mother

At the end of the 1950s, Bowlby aimed to unify ethology and psychoanalysis, to synthesize both approaches into a single framework that explained the nature of a child's attachment to the mother as an instinctual
relationship. At first glance it is hard to understand why Bowlby would be
so enthusiastic about ethology. Lorenz's views on human behavior were
loose extrapolations from his animal studies, based only on analogies and
anecdotes. And Bowlby had already argued in the WHO report that his
views about the significance of mother love in children were conclusively
proved in humans. So what could be gained by finding out how ducks and
geese raise their offspring?

Bowlby's turn to ethology was probably influenced by the mounting
criticism directed during the mid-1950s toward the studies on maternal
deprivation. In addition, he had already been searching for a naturalistic basis to support his views about the mother-child relationship. From
early on in his career, Bowlby was interested in finding a biological foundation for the child's emotional needs. In his first papers he had discussed
animal research on primates and considered this work relevant to explanations of human behavior. He looked at studies of animal behavior to
compare their results with his observations of children.[44] Bowlby always
wrote about the "natural" family unit, emphasized the natural basis of
the mother-child dyad, and portrayed the child's emotional development
as similar to embryological development. During the mid- to late 1950s
he studied ethology and discussed its application to child development in

numerous meetings with researchers such as those who had attended the WHO meetings described in chapter 2.

Furthermore, psychoanalysts had been struggling with the notion of instinct since Freud postulated that human drives originate from a somatic source. Freud said that one day biologists would help psychoanalysts clarify this obscure concept.[45] On encountering ethology, Bowlby believed that day had arrived. He justified his turn to ethology by noting that the need to resort to biology had been "approved," even encouraged, by the founding father:

> In this context it is interesting to reflect on Freud's belief expressed over forty years ago (Freud 1915) that, for a further understanding of instinct, psychology would need to look for help from biology. As a result of developments in the biologically rooted science of ethology I believe the time has now come and that the psychoanalytic theory of instinct can be reformulated.[46]

The next year Bowlby published "The Nature of the Child's Tie to His Mother," where he presented the main thesis for an ethological theory of attachment behavior. In this paper, Bowlby identified two major explanatory frameworks to account for children's attachment to their mothers. Psychoanalysts focused on the infant's gratification and emphasized the mother's role in providing both food and care. He noted that many psychoanalysts of different tendencies supported this theory of secondary drive, including Dorothy Burlingham, Anna Freud, Melanie Klein, Margaret Ribble, Therese Benedek, and René Spitz. However, he saw a discrepancy between their position and their empirical data. He noted that in "each case" they had "observed non-oral social interaction between mother and infant" and, "in describing it, however, each seems to feel a compulsion to give primacy to needs for food and warmth and to suppose that social interaction develops only secondarily and as a result of instrumental learning."[47]

As a major alternative to this widely held psychoanalytic position, he laid out the views of ethologists, "who have never assumed that the only primary drives were those related to physiological needs. On the contrary, all their work has been based on the hypothesis that in animals there are many in-built responses which are comparatively independent of physiological needs and responses, the function of which is to promote social interaction between members of the species." Thus ethologists focused on innate or instinctual social responses, present at birth, not derived by conditioning through the gratification of primary needs like the need for food.[48]

Bowlby then advanced his own theory of component instinctual responses, which he put forward as a synthesis of psychoanalysis and ethology. In his view, attachment behavior integrates several component instinctual responses that at first are independent. The instinctual responses, which mature at different times during the first year of life and develop at different rates, bind the child to the mother and contribute to the reciprocal dynamic of binding mother to child. Five instinctual responses compose attachment behavior. The baby is the active partner in three of them: sucking, clinging, and following. The other two, crying and smiling, serve to "activate maternal behaviour."[49]

Bowlby adopted the ethological, and specifically Lorenzian, conception of instinct. He noted that, rather than using the "cumbersome term 'species-specific behaviour pattern,'" as ethologists did, he would call them "instinctual responses."[50] But Bowlby emphasized that he was using instinct in the ethological sense, not the psychoanalytic sense. He explained that in ethology instincts were "behaviour patterns . . . common to all members of a species and determined in large measure by heredity. They are conceived as the units out of which many of the more complex sequences are built. Once activated the animal of which they form a part seems to be acting with all the blind impulsion with which, as analysts, we are familiar."[51]

He also adopted the concept of social releaser from Lorenz. As an example of how the system of social releasers and instincts works in bonding the child to the mother, Bowlby described the baby's smile: "However activated, as a social releaser of maternal behavior it is powerful. Can we doubt that the more and better an infant smiles the better is he loved and cared for? It is fortunate for their survival that babies are so designed by Nature that they beguile and enslave mothers."[52]

Bowlby thus presented the mother-child dyad as a biological system. In this system, each part complements the other, as designed by evolution. Not only is the child tied to the mother; the mother is tied to the child. Furthermore, he argued that "the nature of the child's tie to his mother" is a natural bond required for the survival of the species.

By appropriating the ethological framework, Bowlby argued for the biological significance of his theory and then claimed support for it from biology:

> I wish to emphasize that it is a main part of my thesis that each of the five
> instinctual responses which I am suggesting underlie the child's tie to his

mother is present because of its survival value. . . . The theory of Component Instinctual Responses, it is claimed, is rooted firmly in biological theory and requires no dynamic which is not plainly explicable in terms of the survival of the species.[53]

As Evelyn Fox Keller has shown in her discussion of molecular biology's reliance on the social authority of physics after World War II, a discipline can benefit from associating itself with another discipline that has higher social status and scientific standing.[54] In this case, by presenting his attachment theory as a synthesis of psychoanalysis and ethology, Bowlby bestowed on it the authority of biological knowledge. In addition, as ethology became very successful after World War II, attachment theory also gained momentum from the popularity of ethology and, specifically, Lorenz's studies of imprinting. As I showed in chapter 2, Lorenz visited the United States on several occasions and his views on parental care in animals received public acclaim.

Bowlby was aware of the advantage conferred on his views by the association with ethology. When he was finishing writing "The Nature of the Child's Tie," in the fall of 1957, he reported from Palo Alto: "As it happens this paper is going to come in for quite a lot of discussion in the next week or two. It is being circulated to one of the seminars and looks like being read more widely. . . . By an unexpected but fortunate coincidence, Konrad Lorenz is going to be here next week for a couple of days. He will stay with us and give some sort of lecture on Tuesday, so ethology and all that seems likely to be in the news."[55]

In the following spring, Bowlby publicized his views in the East Coast. The trip included a visit to Mary Ainsworth in Baltimore, the debate at the *Ladies' Home Journal*, a broadcast on mental care and child care in New York, and more academic visits in Orange Park, New York, and New Haven. He was also hired for five days as a consultant by the National Institute of Mental Health, half-time for the Adult Psychiatry Unit and half-time for the Child Development Studies. Here, he also presented his work on what he called "ETHO-PSYCHO-ANALYSIS." He reported that his ideas were "greeted with some enthusiasm." In addition, he presented "The Nature of the Child's Tie" to the Western New England Psychoanalytic Society. Again, he sent home good news: "The paper was well received, partly perhaps because ethology is not unknown in these parts. Fritz Redlich is an old friend of Konrad Lorenz and has spent time at Konrad's place. So far I have given the paper 4 times in U.S.A. and have had fewer bricks hurled at me in all 4 times together than in London last summer."[56]

Riding on the coattails of ethology's success in the United States, Bowlby's views about the biological foundation of the mother-infant dyad attracted widespread interest. His timing was right. His presentation of the ethological theory of attachment dovetailed nicely with Lorenz's defense of the instinctual nature of human social behavior. Thus, in the United Sates, Bowlby found fertile ground for the idea that ethology could provide a sound framework to understand the attachment of a child to his mother.

But in turning attachment into a biological construct and presenting the mother-child dyad as a natural unit, Bowlby also had to confront one of the most vexing questions for his position: Is the natural mother the only one who can provide adequate child care? After the publication of his WHO report, this issue had been at the center of the social discussions about the role of mothers and the impact of child care. Bowlby had been ambiguous about this. In his 1958 foundational paper, he confronted the issue.

The Power of Natural Love

One of the most controversial issues in Bowlby's writings about the mother-child relationship concerned whether only biological mothers could provide their children with the necessary love and care. Could several people do so? Could fathers?

In his work, Bowlby explored only maternal separation and deprivation. However, sometimes he added a note saying that he was referring to the mother, mother substitute, or mother figure—to the person taking care of the infant, who might not be the biological mother. Aware that this was one of the most controversial implications of his theory, Bowlby was often ambiguous about whether a mother substitute is as good as the natural mother. However, when he situated his views within a biological framework, he was forced to confront the issue: Is the biological mother the one best suited to care for her child?

If evolution had designed the mother-child dyad as a unit, did that imply that biological mothers are the only ones able to provide good care, since they are the ones designed by nature for the role? In his 1958 paper charting the nature of the child's tie to the mother, Bowlby described attachment responses as "mother-oriented," though he said it was evident that this was so "only potentially." Each response could in principle focus on an object other than the mother. But in practice "this is improbable, since all or most of the consummatory stimuli which terminate them habitually come from the mother-figure." Even here, Bowlby talked about a mother "figure," but as he developed his views, his writings indicated that only the mother was designed by nature to provide specific responses to her child's demands.[57]

To clarify his position, Bowlby drew a parallel with the English monarchy:

> It is for this reason that the mother becomes so central a figure in the infant's life. For in healthy development it is towards her that each of the several responses becomes directed, much as each of the subjects of the realm comes to direct his loyalty towards the Queen; and it is in relation to the mother that the several responses become integrated into the complex behaviour which I have termed "attachment behaviour," much as it is in relation to the Sovereign that the components of our constitution become integrated into a working whole.[58]

Thus, not only the laws of nature but the functional rules of a well-ordered society required that authority be invested in a single figure—the queen in society, the mother at home. To ensure social order, the mother thus needs to fulfill her own natural role, as has been proved by the well-functioning designs of nature and society. Bowlby continued:

> We may extend the analogy. It is in the nature of our constitution, as of all others, that sovereignty is vested in a single person. A hierarchy of substitutes is permissible but at the head stands a particular individual. The same is true of the infant. Quite early, by a process of learning, he comes to centre his *instinctual responses* not only on a human figure but *on a particular human figure*. Good mothering from any kind woman ceases to satisfy him—*only his own mother will do*.[59]

In his picture, a hierarchy of authority is necessary for the proper functioning of the natural and the social orders: "The tendency for instinctual responses to be directed towards a particular individual or group of individuals and not promiscuously toward many is one which I believe to be so important and so neglected that it deserves a special term: monotropy."[60]

Bowlby's analogy between the family and the English monarchy is potent. It helped drive home the point that, without a central authority figure at home, the family unit would disintegrate. The mother as queen of the home ensures family stability, much as the British queen ensures social order.

Further, it helped him avoid the charge that he was committing the naturalistic fallacy. Although initially it looks as if Bowlby is claiming monotropy should be a rule of society because it is a law of nature, a second look shows that his argument is more complex. If he said only that mothers are best able to provide good care because they are the ones designed by nature for the role, one could ask, But why should society follow the designs of nature? Moral philosophers have often pointed out the fallacy of going from natural descriptions to moral prescriptions. But what Bowlby said is that not only nature, but also society has shown the functional desirability of monotropy. Thus, in establishing an analogy between the mother-child dyad and the English monarchy, he justified the moral authority of the natural order by appealing to the social order as much as he justified the social order as based on a division of parental roles rooted in the designs of nature.

The turn to biology helped Bowlby legitimize a normative view of motherhood that he had already presented in his WHO report. But turning the child's love for the mother as well as the mother's love for her child into a biological instinct had far-reaching consequences for our understanding of mother love, mothers' work, and gender roles.

The biologizing of love had important implications for mothers. Never before had maternal love been categorized as a child's biological necessity. Nor had the mother been identified as a psychic organizer. In earlier times, a mother could enable or constrain her children's capacities. Mothers could temper, control, and educate their children. But now, according to Bowlby, children have a uniform, universal need for a specific type of mother love, while a mother's feelings determine her children's minds.

Second, although the mother acquires a central role in a child's emotional development, mother love is devalued. In arguing that the mother is designed to fulfill her child's instinctual needs, Bowlby transformed maternal love and care from a personal choice entailing devotion, work, patience, dedication, and not a few renunciations into a natural product of a woman's biological constitution. Furthermore, when maternal feelings are understood as the products of biology, they are removed from the realm of intelligence and freedom, and thus from the realm of behaviors that deserve moral recognition. Bowlby thought he had an upbeat message for mothers: "The normal mother can afford to rely on the prompting of her instincts in the happy knowledge that the tenderness they prompt is what the baby wants."[61] But when Bowlby noted that the normal mother is the unthinking and natural mother who is driven by her instincts, he divested maternal care of choice and, therefore, of moral value.

Third, Bowlby's view of mother love as a child's biological need justified gendered parental roles in a way that had a profound emotional hold on mothers. It is important to grasp not only the logic of an argument but also its emotional force. A mother was now called on to stay home not to fulfill society's desire for social order, but to fulfill her children's natural needs. The emotional power arises because women were not being asked to sacrifice their personal desires for the greater good of the social organism. Early in the Cold War, the rhetoric of containment helped to justify the separation of spheres. But as time went on, even that was unnecessary. If the mother is her child's psychic organizer, she needs to stay home, regardless of whether the times bring war or peace.

Thus, at a crucial period in reassessing parental roles in American society, Bowlby's views about the biological need for mother love represented a strong emotional argument supporting separate parental and gender roles. Although by the 1950s companionate marriage was often touted as an ideal, the different parental roles assigned to men and women made this ideal practically unattainable and socially undesirable. There was an increasing call for fathers to be kind, gentle, and loving to their children, but there was no broad effort to reassess the roles of mother and father. The traditional family unit seemed necessary, as Bowlby and many others pointed out. A father's main role was to provide financial support. As historian Robert L. Griswold has argued, "To support children financially while fostering their sex-role adjustment became the essence of 'maturity,' 'responsibility,' and manhood itself." In the 1950s the father remained the breadwinner, while the mother was responsible for child rearing. The emotional force of appealing to children's biological need for love justified the traditional division of labor with mother at home and father at work.[62]

Conclusion

In the United States Bowlby became an important voice in the scientific and social debates about children's needs and their consequences for the distribution of parental roles. I have argued that the impact of his views about the significance of mother love for child development needs to be understood in the context of the widespread anxiety about changing gender roles and working mothers during the early Cold War years.

By situating discussions about child development in their scientific and social context, I have provided an explanation for the high visibility of Bowlby's work and the corresponding neglect of his critics. Bowlby's ideas about mother love and the implications he drew about the importance of the nuclear family resonated with pervasive social concerns about gender roles in midcentury America. Since American child analysts had already made the role of the mother in helping her child develop the capacity to love into a central public concern, Bowlby was able to gain immediate prominence in ongoing discussions about parental roles. Amid the Cold War emphasis on the benefits of gendered parenting and the nuclear family, the numerous criticisms of the empirical evidence Bowlby used to support the essential role of mother love remained peripheral.

Furthermore, I have argued that Bowlby introduced a crucial new element in the history of justifications for gender roles by claiming that mother love was a biological need for children. Although the significance of mother love had already been part of the history of emotionology in the United States, his appeal to biology to justify the child's innate need for the mother secured it to a new scientific foundation, one that held a tremendous emotional power.

Understanding Bowlby's interpretation of "the *nature* of the child's tie to his mother" and interpreting its scientific and cultural reception during the 1950s cannot be done without locating its place in the debate about the mother's *social* tie to her child. Bowlby appealed to biology to claim that his position was based on scientific knowledge independent of social views and contingent historical factors. But the development and impact of scientific pronouncements about mother love have to be understood in the context of the changing concerns about women's participation in the workforce and the recurring debate about the distribution of parental roles. Questions about the nature of mother love and children's needs remained entangled with the question "Who should rear the children and how?"

PART 2

Challenging Instincts

CHAPTER FOUR

Against Evolutionary Determinism

The Role of Ontogeny in Behavior

Introduction

In the late 1950s, John Bowlby presented a synthesis of psychoanalytic and ethological ideas in support of his views about children's innate need for maternal love. Yet as he developed his views on attachment behavior, the main concepts he imported from ethology—instinct and imprinting—received widespread criticism among animal researchers.

In 1953, the year Bowlby first appealed to Lorenz's work in print, American comparative psychologist Daniel Lehrman thoroughly critiqued the ethological program of identifying instinctive behavior. He argued that deprivation experiments cannot identify innate behavior, as Lorenz claimed. According to Lehrman, a researcher will be unable to determine whether a trait or behavior is more or less affected by experience and environmental factors without knowing its developmental history. He emphasized the need to understand the physiology and ontogenetic development of an organism's behavior in its ecological and social context. Lehrman further criticized ethologists for extrapolating from their animal research to human conduct. Burkhardt, Griffiths, and Vicedo have analyzed several aspects of this controversy and have shown the significant influence Lehrman's views had on other researchers. In this chapter, I focus on Lehrman's rejection of instincts and show his alternative understanding of maternal care in birds.[1]

Further, I show that other animal researchers rejected several core tenets of ethology. Comparative psychologists Frank Beach, T. C. Schneirla,
and Jay Rosenblatt stressed that isolation experiments cannot separate
innate and learned behaviors. British animal researcher Robert Hinde exposed the weaknesses in Lorenz's hydraulic model of instinctual energy
and the concept of drive. Several researchers demonstrated that the way
imprinting works in some birds could not be generalized to other species.
They questioned the existence of critical periods for developing filial and
sexual responses and thus whether imprinting was irreversible, as Lorenz
had claimed.

By examining the critical analyses of Lorenz's framework, I show that
by the mid-1960s many students of animal behavior had moved beyond
his views about instincts, drives, and imprinting. To defend the validity of
his program, Lorenz often accused his critics of denying the biological
basis of human behavior for ideological reasons or rejecting instincts because they were ignorant about biology. Nevertheless, many animal researchers, including Tinbergen, concluded that the concepts originally employed in ethology, such as drive, instinct, and innateness, were too broad
and imprecise to advance the study of animal behavior. They increasingly
rejected Lorenz's conception of instinctive behavior and his views about
the determinant effect of early experiences on an animal's adult behavior.

Daniel Lehrman: Against Konrad Lorenz's Theory of Instincts

Daniel Lehrman (1919–1972) was among the first American scientists
to be well acquainted with the new ethological approach to animal behavior. While only a teenager, he carried out research with herpetolo-
gist G. Kingsley Noble at the American Museum of Natural History
(AMNH) in New York. Here he interacted with leading biologists. Like
Lorenz and Tinbergen, Lehrman studied birds, and he shared their love
for bird-watching. He was a superb ornithologist, having played hooky
from school on many spring days to watch birds in Central Park, often in
the company of world-class authorities like evolutionary biologist Ernst
Mayr, who worked at the AMNH at the time. Tinbergen met the "shy,
keen, and thin-as-a-beanstalk" nineteen-year-old Lehrman during his first
visit to the United States in 1938.[2] In 1941 Lehrman, who at twenty-two
had a solid command of German, reviewed a recent paper by Lorenz and
offered interested readers copies of his own English translation.[3] World

War II sidetracked his studies. He spent it in Europe helping to decipher coded messages. On his return, he pursued a PhD with comparative psychologist T. C. Schneirla.[4]

After earning his doctorate at the University of Michigan, Schneirla (1902–1968) had joined the Department of Psychology at New York University in 1927. In 1943 he became associate curator and later curator at the AMNH's department of animal behavior. Schneirla was one of the earliest American experimental psychologists to venture beyond the comforts of the lab. Starting in the early 1930s, he went on research field trips to Mexico, Panama, and Trinidad. He further departed from well-trodden paths in psychology by studying not the conditioned behavior of rats, but the natural conduct of army ants.[5] His work focused on the interrelations among behavior, environment, and physiological processes. He always emphasized the complexity of such interrelations and the difficulty of separating the causes of behavior into independent internal and external factors, genetic and environmental causes. His student Lehrman completely agreed.

In the United States, Schneirla and Lehrman became the main representatives of an approach to the study of animal behavior that differed from the American behaviorist psychology prevalent at the time. Although comparative psychologists were sometimes called animal behaviorists, the approach taken by Schneirla and his students had nothing to do with behaviorism of the Watson-Skinner type that explained behavior as a result of conditioned learning. Schneirla and Lehrman were sympathetic to the ethological goal of studying a variety of animals in their natural environments as well as under controlled laboratory conditions, but they also challenged some ethological ideas.

Schneirla argued that Tinbergen and Lorenz paid too little attention to the role of experience in the development of behavior and that there were no reliable criteria for identifying innate behaviors. But Schneirla embedded these criticisms in presentations of his own complex research, where the sharpness of his analysis was often lost in dense prose, empirical detail, and epistemological nuance.[6] As a consequence, the most visible critical review of ethology would be the one prepared by his student. The title of Lehrman's 1953 paper made the objective clear: "A Critique of Konrad Lorenz's Theory of Instinctive Behavior."[7]

What's the use of calling a behavior "innate" or "instinctive"? Lehrman asked. He admitted that some behaviors are characteristic of one species, stereotyped, and constant in form. In addition, they appear in animals

FIGURE 4.1. Daniel Sanford Lehrman, ca. 1965. Courtesy of Naomi May Miller.

raised in isolation from others, animals that had not previously practiced them. However, Lehrman doubted much insight could be gained by lumping all behaviors of all species that show those characteristics under the umbrella of the innate or instinctive, as Lorenz proposed. Calling a behavior instinctive, Lehrman claimed, does not explain anything about it.[8]

Ethologists also asserted that the behaviors they identified as instinctive are inherited, but Lehrman argued that such a characterization smacked of discredited preformationist and teleological thinking. As he put it, "Lorenz undoubtedly does not think that the zygote contains the instinctive act in miniature, or that the gene is the equivalent of an entelechy which purposefully and continuously tries to push the organism's development in a particular direction." Yet it seemed to him that Lorenz made those assumptions when he claimed that some behavior patterns are "inherited." Furthermore, the idea that innate behavior develops independent of external influences assumed that ontogenetic development is a simple and directed unfolding of what is already present in the zygote. For Lehrman, however, individual development does not follow such a preestablished, deterministic path.[9]

Lehrman, like his mentor Schneirla, saw behavior as the result of a developmental process in which interactions between an organism and its environment affect the course of the process at all stages. Here he was making an ontological claim, an assertion about the way things are: Behavior, he said, is not set at conception. It is not determined by an individual's hereditary makeup as configured by its evolutionary history. This ontological claim also led him to a specific epistemological stand, a view about what scientists can know about behavior and how. Because of the early and continuous interaction in the development of behavior between internal and external factors, environmental and physiological inputs, and experiences and organic conditions, Lehrman doubted that researchers could separate the innate components of behavior from the learned ones. More specifically, he denied that deprivation experiments—the method Lorenz proposed—could do it. Lehrman contended that a researcher could never be sure the individual animal under study had been isolated from all relevant factors that affect the development and performance of any given behavior. He wrote, "It must be realized that an animal raised in isolation from fellow-members of his species *is not necessarily isolated from the effect of processes and events which contribute to the development of any particular behavior pattern.*" In Lehrman's view, an isolation experiment can provide, at most, "a negative indication that certain specified environmental factors probably are not directly involved in the genesis of a particular behavior." But the isolation experiment does not provide any information about the developmental process that leads to the behavior. Therefore a researcher cannot determine whether a behavior appears independently of any experience or environmental influence.[10]

Finally, Lehrman criticized the ethologists' extrapolations from animal behavior to human conduct. In a section titled "The Human Level," Lehrman questioned the value of juxtaposing behaviors from different species under the same rubric. He wondered, for example, what "valid explanatory purpose" could be served by describing both a bird's search for a nest site and a man's search for a house as examples of "searching behavior," as Tinbergen had done. As Lehrman saw it, the only basis for this juxtaposition was that "*the outcomes are similar from the human point of view*": the bird gets a nest, and the man gets a house. But what does one really learn about the bird or the man by including their conduct in the common category of "searching behavior" and then hypothesizing that it is instinctive? Little or nothing, Lehrman thought. He also cited problematic examples from Lorenz's ethological explanations of human behavior, such as his 1940 interpretation of "the relative attractiveness to women of several breeds of dog in terms of the degree to which they fit the innate perceptual pattern releasing instinctive maternal behavior in the human individual!" Lehrman believed such a statement was an unwarranted extrapolation to human beings of research in animal perception of visual shapes.[11]

Lehrman further pointed out that some of the social policy measures Lorenz had drawn from his theory were not only scientifically unsound, but also racist. For example, Lorenz had argued that unrestricted breeding in humans leads to a degeneration of the innate releasing mechanisms connected to mating behavior. In Lorenz's opinion, this was disastrous because people no longer were prevented from mating with individuals from other races and with those displaying "degenerate" social instincts and features. For Lorenz, this provided a scientific reason to support social prohibitions preventing some groups from interbreeding. He also presented these views in a 1940 discussion of the scientific justification for legal restrictions against marriage between Germans and non-Germans.[12] Since Lorenz saw these political ideas as following from his biological views, Lehrman thought it was appropriate to challenge his sociopolitical views as unwarranted conclusions about the nature of instincts and their degeneration in humans.

"Lehrman's paper sent a shock wave through the ethological community," claims historian of science Richard Burkhardt. Ethologists were forced to address Lehrman's criticism of their concept of innateness and their reliance on isolation experiments to identify the innate components of behavior. After all, these were main tenets of the fledging science of ethology. In the paper he presented at the 1949 Cambridge symposium, Lorenz had staked the identity of the new field on the existence of instincts and on researchers' ability to identify them. Tinbergen told Lorenz about the stir caused by Lehrman's paper in the language they understood best: "The community of ethologists was humming like a disturbed bee-hive." In the following years, Lehrman's criticism became the center of debate in several meetings of the community of animal researchers.[13]

Schneirla, Lorenz, and Lehrman discussed the concept of instinct at a 1954 conference in Paris also attended by leading British geneticist J. B. S. Haldane and Tinbergen's student Desmond Morris, among others. Lorenz complained that his critics were not "getting the facts." He noted that he was not accusing them of suppressing the empirical evidence; rather, he suggested that "semantic difficulties may also have contributed to this failure." This seemed like an encouraging beginning, but Lorenz did not remain conciliatory for long. Soon he accused Lehrman of having made "the wildest caricature of scientific truth" by rejecting the facts "for purely dogmatic reasons."[14] In this way he presented himself as a brave scientist who was simply defending the facts, whereas his opponents were moved by ideological reasons.

Elsewhere I have examined the controversy over Lehrman's critique, probing how the different participants used the categories of the "semantic," the "empirical," and the "ideological" as weapons. There were indeed empirical facts to be interpreted, semantic issues to be clarified, and important political differences in the debate. As I mentioned in chapter 2, Lorenz had been a member of the Nazi party. Lehrman was a leftist, and he was also Jewish. Although there can be little doubt that their different backgrounds and social beliefs influenced their scientific concerns, their differences about how to explain animal behavior cannot be reduced to politics. Lorenz and Lehrman came from different research traditions, and their views were firmly grounded in their research programs and results.[15]

Thus it is important to see that Lehrman had a different vision of how to explain animal behavior, resulting from his experimental and observational work. For our purposes here, I focus on his views on parental care.

Behavior without Predetermination: Lehrman on Maternal Care

The difference between Lorenz's and Lehrman's models of animal behavior is fundamental and led to different research programs. In Lorenz's model, behavior results from the unfolding of a predetermined plan set forth by evolution. An organism is moved from behind, so to speak. The hereditary makeup designed by its evolutionary history determines its behavior. In Lehrman's model, on the other hand, behavior results from continuous interaction between an organism and its environment. The organism is also a product of an evolutionary history, but that history does not predetermine its behavior. Within certain limits, the organism's par-

ticular environment and experiences profoundly shape its behavior. So
the organism is not determined by the past. Instead, what it becomes depends on the combined influences of the past and present. We can appreciate Lorenz's and Lehrman's different interpretations of living organisms and their implications for research by comparing what they said
about maternal behavior in birds.

For Lorenz, the inborn schema or innate releasing mechanism of the
female bird determines how she parents. Her behavior is an automatic response to the behavior of the infant bird. As Lorenz put it, the lock determines the form of the key. There is no room for freedom. If the mother
bird performs behaviors outside the fixed-action patterns of her species, the key will not open the lock, and the infant will not develop the
standard set of behaviors characteristic of its species. Evolution has designed the mother bird to act in a predetermined way. That is why the
Cairina mother rescued the strange chick from the experimenter's hand,
even though she immediately killed it when it tried to mix with her own
offspring. The *Cairina* mother is programmed to save an infant. That is
also why Lorenz's infant daughter went running toward the doll. For Lorenz, females are programmed to care for a cute baby, and babies are programmed to perform a set of cues that release maternal behavior. In this
model of behavior, development is understood only as the unfolding of
what is already programmed by the evolutionary history of a species. The
behavior of any particular organism is just a token of the type, and the
type is already set. That is also why for Lorenz variations are pathological.
In consequence, the ethologist's task is to uncover the type, or the innate
behavioral program already designed by evolution.

In Lehrman's alternative model, an organism's genetic inheritance
does not determine its behavior. Both in utero and after birth, the organism develops in a specific context, which differs from one organism to
the next, even within the same species and the same ecological habitat.
Within its particular context, an organism undergoes a series of events
and experiences that shape its behavioral repertoire. In addition, its behavior has an effect on its surroundings, and those changes in turn will affect its future conduct. Development is thus a dynamic, creative, and individual process with specific contingencies and unique historical twists that
shape the animal's behavior. Because each organism develops in a unique
historical sequence, studying development is essential for understanding
its behavior. It is not sufficient to know that an organism is a token of a
type. In this model, variations are not irrelevant but are an intrinsic part

of a living organism. They are not considered pathological deviations of a species's standard and fixed behavioral repertoire.

Lehrman's research on the interaction between hormones and behavior critically influenced his views on the open-ended character of behavioral development. Frank Beach had carried out extensive work showing how hormones affect behavior. Lehrman was among the first investigators to show that behavior affects hormonal processes as well: an organism's biology affects its behavior, but its behavior also affects its biology. So biology does not predetermine behavior. Biology sets limits by defining the possibilities, but an organism's behavior is also shaped by its specific experiences and environment and by its own doings.

So what does maternal care look like from Lehrman's perspective? Lehrman did most of his experiments on the ringdove (*Streptopelia risoria*), a buff bird with a black collar, belonging to the dove family. The breeding behavior of ringdoves goes through a series of stages: courtship, mating, nest building, incubation, brooding the young, and feeding them. When a male and female are together, they will repeat this cycle about every five weeks. But the female will not produce eggs unless she sees a male. This visual stimulation is necessary before the pituitary gland secretes the hormones required to start her cycle. Both parents incubate the eggs. During incubation, they also secrete crop milk and store it in a pouch in the wall of the esophagus, where food is normally stored before it enters the stomach. After a hungry squab raises and moves its head in the nest, the parent grasps the baby's bill, and the squab in turn thrusts its head into the parent's throat, where the milk will be regurgitated.[16]

Is the parents' behavior predetermined, or does experience play a role in the way they feed their babies? Lehrman had observed that experienced parents tapped the squab to stimulate its head movements, whereas in first-time parents the squab always initiated the movements to start eating. So he postulated that parents learn to arouse the squab at feeding time. To test this, he took two groups of ringdoves, one group with experience feeding the young and the other without experience. He injected both groups with prolactin, a pituitary hormone that stimulates secretion of crop milk. All the doves developed crop milk and were given squabs from other families. The inexperienced birds did not know what to do and could not relieve the tension in their crops by feeding the babies. The control birds that had not been injected with prolactin attacked the squabs. Only the experienced birds approached the babies, pecked at their bills to initiate the head movements, and started to feed them. But when Lehrman anesthetized the crops of these experienced birds, they did not approach to feed the squabs, although they seemed to feel tension caused by the crop milk. So the experience of breeding had "an effect on the readiness of the animal to respond to the hormone by performing the feeding act." Physiological and behavioral changes were deeply intertwined.[17]

For Lehrman, an organism does not "possess" a core set of innate behaviors to which experience "adds" something. The behavior is not ready-made. Rather, it is the result of the interaction of biology, environment, and experience. That is why isolation experiments are not adequate to elucidate the source of an animal's conduct. Since its behavior results from a series of complex interactions beginning at birth, Lehrman argued that simply isolating an animal from one specific factor cannot reveal whether the behavior is instinctive.

Lehrman's conception of an organism constituted a radical ontological challenge to Lorenz's vision of animal behavior—a fundamental disagreement about the ways things are in nature. In addition, his critique of isolation experiments was an epistemological challenge to Lorenz's program for identifying instinctive behavior. On these points, many researchers came to agree with Lehrman.

The Impossibility of Isolating the Innate

Leading experts on behavior, including the Canadian comparative psychologist Donald O. Hebb, Frank Beach, and even Niko Tinbergen accepted Lehrman's position that isolation experiments cannot decisively identify "innate" behavior and that Lorenz's concept of instinct should be rejected.

Hebb's case is interesting because initially he had been skeptical about Lehrman's criticism of ethology. In 1949 Lehrman wrote asking Hebb to comment on an earlier version of his 1953 critique of Lorenz's theory of instincts. Hebb did not know Lehrman, so he first wrote to Beach to find out whether he should be honest with Lehrman: "I don't like the paper at all, on a number of points." Beach did not mince words: "You should come down with a heavy hand on any weak points." Although Hebb told Lehrman he approved "the main thesis of the paper, which is adequately and effectively presented in the earlier parts," he thought the paper also revealed an "*a priori* bias, against 'instinct' ideas." Hebb stood on the other side: "I should say at once that I am biased the other way, and you must decide for yourself how far to discount my criticisms on that ground." Finally, Hebb recommended to "Dr. Lehrman" (although Lehrman was still doing his doctoral research) that he postpone publication, organize the paper better, and include some additional literature.[18]

Four years later, when Lehrman's paper appeared in print, Hebb wrote to Lehrman again, telling him that their positions were closer than he had originally realized. He now thought that the ethological goal of first studying innate behavior was misguided and fruitless. In his 1953 paper "Heredity and Environment in Mammalian Behaviour," Hebb still asserted that his bias was "on the nativistic side," but he faulted the ethological approach for identifying as innate whatever is not learned, and for assuming that "unlearned" is the same as "genetically-determined." Hebb also argued that there are not "two kinds of control of behaviour," one for innate, another for learned behaviors. Thus, he concluded, "the term 'instinct,' implying a mechanism or neural process independent of environmental factors, and distinct from the neural processes into which learning enters, is a completely misleading term and should be abandoned."[19]

The case of Frank Beach is also significant because Lorenz reported that, after seeing Lorenz's films, Beach had dropped all objections to ethology and said in a small voice: "You know I did not believe a word of it and now I believe everything."[20] However, this sounds like a self-serving account. It was unlikely that Beach spoke in a small voice, since he was notorious in the scientific community for his boisterous personality. More important, right about that time Beach published "The Descent of Instinct" (1955), and his views there were hardly in agreement with Lorenz's program to study instincts. Interestingly, in this paper Beach did not cite Lorenz, Lehrman, Schneirla, or Tinbergen, though he knew all of them personally and was well aware of their debates on this issue. But after a brief historical account of the instinct concept, Beach engaged in a critical analysis of instincts "as complex, unlearned patterns of behavior," which was the ethological definition of instincts.[21]

Beach presented "three serious criticisms" against the "current treatment of the problem of instinctive behavior." First, he said, psychologists knew little about many of the behaviors often classified as instinctive. Second, despite their ignorance about those behaviors, there was "strong pressure toward premature categorization of the as yet unanalyzed patterns of reaction." Third, he found that the definition of instincts as unlearned behavior forced psychologists to use a "two-class system" to classify complex behavior. Such a system would imply that behavior is de-

termined either by learning or by heredity. But Beach considered that di-
chotomy "entirely unjustified."[22]

For Beach, scientists needed to clarify two main things: the relation
between genes and behavior, and the ontogeny of behavior. Regarding
the first, he noted that the work already performed cast "doubt upon the
validity of any classificatory system which describes one type of behavior as genetically determined and another as experientially determined."
He furthermore asserted that no behavior could be understood without
knowing its ontogenesis. Beach predicted that when both the role of heredity and the process of ontogenesis were better known, "the concept of
instinct would disappear, to be replaced by scientifically valid and useful
explanations."[23]

Perhaps to avoid antagonizing other researchers or to prevent further
polarization between ethologists and comparative psychologists, Beach
did not present himself as taking Lehrman's side in his controversy with
Lorenz. But he said pretty much what Lehrman had said earlier, that calling a behavior instinctive does not explain anything about it and that investigators needed to focus on the ontogeny of behavior. In fact, he noted
that "the degree of assurance with which instincts are attributed to a given
species is inversely related to the extent to which that species has been
studied, particularly from the developmental point of view."[24]

Nor did Beach find Lorenz's favorite method of isolation experiments
helpful in the search for instincts. Ten years later, in a 1965 review of work
relevant to this issue, Beach agreed with Schneirla and Lehrman that to
report simply that an animal had been reared "in isolation" was meaningless. Beach summarized the results from two studies in his area of expertise—reproduction—to illustrate this point. He reported on the 1955
study by Elliot Valenstein, Walter Riss, and William C. Young with male
guinea pigs reared in separate cages. As adults these guinea pigs copu-
lated less effectively than males reared with conspecifics. Normally, prepubertal guinea pigs mount cagemates. Thus researchers first thought that
preventing this had affected normal development of males' adult copulatory patterns. However, Arnold A. Gerall later published another study
with intriguing results. Control male guinea pigs raised in individual compartments separated by solid partitions showed deficient copulatory patterns as adults. Yet male guinea pigs reared in individual compartments
but separated from conspecifics only by coarse wire mesh exhibited normal copulatory behavior in adulthood. Thus it seemed that seeing the
other individuals was enough to ensure normal sexual activity. They didn't
need to practice mounting in their prepubertal stage.[25]

Beach also reported that Lester Aronson, at the AMNH, found similar effects in fish. Aronson first noted that mature pairs of *Tilapia* placed in aquariums with bare slate floors showed no spawning activity. Then he covered the floor with gravel and covered the gravel with clear glass. In this second setup, although the fish could not touch the gravel, most pairs selected a nest location, executed responses that ordinarily constitute nest excavation, and carried out all the patterns of normal spawning.[26]

These results led Beach to conclude, "Obviously we are as yet unprepared to identify the crucial deprivation involved, but no simple reference to 'prevention of practice' seems applicable." Beach went on: "To further complicate the picture there are definite interspecific differences within the same order. In contrast to guinea pigs, male rats reared in individual cages with solid walls copulate quite effectively at the first opportunity offered in adulthood," as his own work had demonstrated.[27] The general point was that if an organism performs a specific action well without prior experience or training in that specific behavior, that alone does not prove the behavior is innate, since competence to perform it may have been acquired in other ways.

Throughout the 1960s Schneirla and another of his most successful students, Jay Rosenblatt, also challenged the efficacy of isolation experiments in identifying instinctive behavior. They argued that isolation experiments could not be interpreted without reference to longitudinal investigations of the normal behavior of an animal species. As Lehrman had emphasized over the years, studies of ontogeny are not incidental but are essential to Lorenz's goal of finding out which behaviors can develop without experience. Without such studies, one cannot even interpret isolation experiments.

The other major founder of ethology, Niko Tinbergen, acknowledged that the study of ontogeny needed to be a central part of any science of animal behavior. Very soon after Lehrman's 1953 paper and the discussion about instincts in Paris, some American comparative psychologists and European ethological researchers met again in Ithaca, New York, at a conference organized by the Josiah Macy Jr. Foundation. Lorenz, Tinbergen, Schneirla, Lehrman, Mayr, anthropologist Margaret Mead, and psychoanalyst Erik Erikson, among others, attended this invitation-only affair. The interdisciplinary group attests to the perception that the issues under discussion were important for other areas of biology and for the social sciences.

From the start, Tinbergen was much more conciliatory than Lorenz had been in Paris. When the participants explained briefly why they were attending, Tinbergen said:

My presence here is due mainly to Dr. Lehrman. His paper in the Quarterly
Review of Biology was most stimulating, in an ambivalent way. Here was a
good example of a phenomenon: Although a great number of our facts are correct, we (at least I) had gone too far beyond the facts. I suddenly saw, when Dr.
Lehrman's paper arrived, that we ought to start anew from the many facts both
of us had, and that we differed chiefly over interpretation.[28]

Hoping to reach common ground, Tinbergen was even willing to re-
nounce a term—and the object it referred to—that was central to the
original ethological program—the innate. He noted: "Let me say right at
the beginning, that I admit that we must drop this use of the word 'innate'
for the reasons already given, namely, that the word can be applied only
to differences, not to characters, and also because tests such as ours exclude only part of all possible environmental influences." Tinbergen left
no doubt on this point: "The quarrel about the use of the term 'innate' can
now be forgotten. I have accepted Dr. Lehrman's criticism on this point,
and I believe we can build constructively from here on."[29] The discussion
in this meeting and later interactions between students of animal behavior convinced Tinbergen that the field would move forward more fruit-
fully by integrating the perspectives and insights of both the ethologists
and the comparative psychologists.

Almost a decade later, in his classic 1963 paper "On Aims and Methods
of Ethology," Tinbergen drifted even further from the original Lorenzian
program. Concerning the concept of innateness, Tinbergen now maintained that "it may not be superfluous to stress that the recognition of the
existence of many species-specific behavior characters does not necessarily imply that all these characters are 'innate' in the sense of ontogeneti-
cally wholly independent of the environment." He found no justification
for "lumping all examples for which not all environmental influences have
been eliminated into a class called innate, thus suggesting a positive statement where merely a negative statement would be in order." He concluded
that to apply "the adjective 'innate' to *behaviour* characters, and to do this
on the basis of eliminations of different kinds is heuristically harmful."[30]

Tinbergen even found "unhelpful" Lorenz's position that the innate
and learned components of a behavior are intercalated. That conception,
explained Tinbergen, led to the questionable view that if only one could
split up behavioral chains into smaller and smaller components, at some
point one would be able to separate the innate components of the chain
from the acquired ones. For this reason, Tinbergen rejected his earlier ap-

proach to the study of behavior. It was not only that the ethologist's "eval-uation of the part played by internal determinants may have been on the optimistic side," but that the original goal of ethology was misconceived: "I cannot see how, in view of such facts, it can be fruitful to look for innate and learned *components*, however small."[31]

Like Beach in 1955, Tinbergen did not mention Schneirla or Lehrman, but his 1963 paper publicly acknowledged the validity of their criticism. Tinbergen then presented a modified program for ethology, which would address four "whys" about behavior: its immediate causation; its develop-ment; its consequences and function; and its evolution. He now agreed with the Americans that behavior could not be understood without pay-ing attention to development and without integrating the knowledge of both ontogenesis and phylogenesis in a specific context.

In sum, a thorough discussion of Lorenz's views about behavior and in-stincts followed Lehrman's critique. Many students of behavior rejected Lorenz's conception of instinct, which he had placed at the core of his ethological program. At the same time, other researchers criticized other central concepts and tenets of Lorenz's original program for ethology.

Hinde against Drives

Lorenz and Tinbergen also used the concept of drive, and they presented models of internal energy accumulation and release, as we saw in chap-ter 2. In the late 1950s, Cambridge animal researcher Robert Hinde pub-lished a series of articles disputing the concept of drive and the energy models underlying it.

Hinde first briefly described Lorenz's and Tinbergen's models of drives. Lorenz's vision of instincts drew on the model of a hydraulic reservoir. He postulated a particular "action-specific energy" for each instinctive act. This energy is accumulated as in a reservoir with a spring valve. Tinber-gen had developed a "hierarchical system of centres," in which he likewise postulated the existence of "motivational impulses" that accumulate in the nervous centers. Those impulses are held in check by a "block," which, like the valve in Lorenz's model, will be removed by the action of internal and external releasing mechanisms. Lorenz had presented his model as a heuristic tool, whereas Tinbergen thought of his model more as a provi-sional representation of the mechanisms of the nervous system involved in instinctive actions.[32]

Hinde argued that the many ambiguities in the term drive were stultifying to research. He identified three main meanings in the literature. One use of "drive" referred to a central state "caused" by hormones, stimuli, and so on, which itself "causes" the animal to act in a particular way. When ethologists used the term drive in this sense, Hinde said, they seemed to be assuming the presence of "existence postulates," as when they said that scientists claimed the drive was "being expended," "having impetus," "flowing," or being "discarded," "thwarted," and "sparking over." In this sense, some ethologists used drive as synonymous with Lorenz's "action-specific energy," one of the components of his hydraulic model of instincts. But others seemed to refer to a mechanism, as when they said the drive was being "activated." Citing specific uses of the word by major ethologists, Hinde considered any attempt to clarify the nature of the phenomena ethologists were talking about when using the concept of drive a "hopeless task." He asked: "How is one to form a picture of an entity which, in addition to being 'discharged,' 'thwarted,' etc., may be 'satisfied' or 'unsatisfied,' is 'consumed by' movements which are 'fed' by it, which is 'obeyed' by the animal, may or may not be 'generalised,' has appetitive behaviour 'belonging' to it, may 'conflict' or 'compete' with other drives, and may even 'lack coordination' with the 'cognitive element'?"[33]

Some ethologists, Hinde noted, used "drive" in a second sense, to refer to "all causal factors other than those received through exteroceptors." And some researchers were using it in a third sense, to refer to "*all* the causal factors influencing the behaviour, whether internal or external.[34] The concept seemed so imprecise as to be problematic.

Hinde found other concepts closely related to the hydraulic model, such as innate releasing mechanism and "displacement activity," problematic as well. What exactly was displaced? "Sometimes it is the 'nervous energy,' sometimes the 'impulses,' sometimes the 'behaviour' (e.g. incubation), sometimes the 'nervous system' and sometimes even the 'fish' which is said to 'spark over.'" Hinde did not deny the existence of displacement activities. However, he found that discussing them in terms of "sparking over" concealed "the inability of the hydraulic model to explain how one set of causal factors seems to produce behaviour 'normally' dependent on another set, and gives a spurious appearance of uniformity to a heterogeneous class of activities."[35] Cast in a friendly tone recognizing that the hydraulic model of instincts and related constructs such as drives had stimulated research on behavior, Hinde's critique nevertheless made clear that at present they were obstructing progress in the field. Using those constructs gave the appearance of understanding mechanisms and processes in animal behavior that were explained only through imprecise models and analogies.

In his 1959 paper "Unitary Drives," Hinde again emphasized that drive concepts oversimplified the causal mechanisms underlying behavior.[36] He also noted that drives were taken as units with a hypothetical physiological reality. However, referring to a drive hindered the search for that physiological reality. There is no single energy or single drive underlying all species-specific behaviors, Hinde noted. Lorenz had recognized this last point and thus had postulated one type of energy for each instinctive behavior. But then what explanatory power did the concept of drive have?

For Lorenz these criticisms marked the end of Hinde as an official "ethologist." When, years earlier, Lorenz had first heard Hinde give a presentation, he was excited about the bright young man's possible contributions to the field. His premonition was correct, since Hinde went on to a brilliant career as an animal researcher at the Ornithological Field Station in Madingley and the Department of Zoology in Cambridge. He worked with birds and then monkeys, and he trained many of the world's best students of animal behavior, including Jane Goodall and Dian Fossey. But after Hinde published his critique of energy models of motivation, Lorenz ignored him and his studies. In private correspondence with other animal researchers, he also made disparaging remarks about Hinde's work.[37] Lorenz treated the field as his personal creation and viewed any criticism as disloyalty. In addition, Hinde had become a close friend of Lehrman, which probably did not help his status in Lorenz's eyes.

Hinde's condemnation of the concept of drive had much in common with Lehrman's criticism of the concept of "innate" behavior. Lehrman had asked: What is to be gained by placing many different behaviors under the same umbrella of the innate? He thought doing so hindered the search for the various physiological, genetic, and neurological processes involved in the ontogenetic development of each behavior, undermining the examination of how those processes were related to the specific environmental and experiential factors influencing each behavior. Similarly, Hinde now asked: What is to be gained by calling the motivational factors that play a role in a diversity of behaviors "drives"? Not much, he concluded. Hinde, like Lehrman, zeroed in on the inconsistencies and lacunae in Lorenz's suggestive but vague concepts. Instincts and drives, they

argued, did not explain anything about an animal's behavior. Soon another central concept in Lorenz's theoretical framework—imprinting—
also came under extensive scrutiny.

Critique of Imprinting

During the 1950s and 1960s, several studies revealed that imprinting is not
the simplistic mechanism Lorenz had first suggested. For Lorenz several
special features characterized the process: it takes place during a critical
period, it is independent from learning, it is irreversible, and it has a determinant impact on adult behavior. Specifically, Lorenz claimed that in
some species filial imprinting determines the adult bird's sexual prefer-
ences. All these characteristics were related to each other.

Lorenz's belief that imprinting takes place during a critical period was
closely connected to his view that it has nothing to do with learning and
that it is irreversible. For Lorenz imprinting is a quick process, circumscribed within clear time boundaries. However, research on a variety
of animals showed that though there are some sensitive periods for imprinting, for most species there are no critical periods in the sense of a
clearly defined time after which imprinting cannot take place, as Lorenz
had claimed. Researchers also showed that in most cases imprinting also
depends on practice and on length of exposure to the object of imprinting. For example, in their study of the social development of the cat, T. C.
Schneirla and Jay Rosenblatt did not find that any particular period was
the key to future behavior. Rather, there was an interrelation between the
stages in the life of the animal. They argued that "striking changes in the
essential progression are grounded not only in the growth-dependent processes of maturation but also, at the same time, in opportunities for experience and learning arising in the standard female-litter situation." Their
conception of social ontogeny stressed "not just one or a few chronologi-
cally marked changes in the behavior pattern, but rather indicates that
normally each age period is crucial for the development of particular aspects in a complex progressive pattern of adjustment."[38] At Yale University, Julian Jaynes carried out a series of studies with domestic neonate
chicks in an apparatus to test whether they would develop a following
response toward a cardboard object in order to analyze different aspects
of imprinting. Among other results, he found that a critical period for acquiring filial responses exists in that species, but he also found that prac-

tice plays an important role in the retention of imprinting.[39] This contradicted Lorenz's assertion that imprinting has nothing to do with learning.
Further, if practice and learning influence the process, they could also be
responsible for the animal's behavior at later stages.

Did the effects of early infancy remain in the adult life of the bird because of the impact of imprinting, or were they due to subsequent reinforcement? Geese imprinted on Lorenz were not interested in conspecifics later in life. But was that because Lorenz was the first living object they
had encountered after hatching or because they had remained in his company throughout their adulthood? It was impossible to tell. As early as
1955 Hinde pointed out, "No really adequate experiments seem to have
been done here, however, for in all cases known to the writer the birds
were more or less continually in the presence of man throughout their
pre-adult life: the attachment of the sexual response to man, though perhaps influenced indirectly by the early experience, could have been due to
a later learning process."[40]

The following year W. H. Thorpe, a strong supporter of Lorenz, raised
the same point: "Since special precautions against reinforcement by subsequent conditioning were not taken, this must have played a large, perhaps very large, part in the later course and strength of the phenomenon."[41] Another researcher, Howard Moltz, reached similar conclusions:
"Since imprinting, *uncontaminated by reward learning*, has not been
shown to affect directly any other response system of the adult animal,
the possibility that it affects adult personality remains unexplored."[42]

Although Lorenz asserted that the effects of imprinting are irreversible, a number of studies showed that, in many species, animals imprinted
on the wrong object or subject could nevertheless develop courting and
mating behavior toward the right members of their species. That is, the
effect of imprinting on mate choice was neither determinant nor irreversible.[43] Further research was needed to clarify the relation between the following response, filial imprinting in infancy, and adult sexual responses.

In addition, researchers disagreed about how much one could generalize from results obtained in one species. Even in birds, not all species
show the same ability to imprint, and even within the same species there
are differences. University of Chicago comparative psychologist Eckhard
Hess, a friend and follower of Lorenz and one of the foremost students
of imprinting, showed that Vantress broiler chicks are good imprinters,
but white leghorn chicks are not. Individual differences within a breed
are also common and can further vary according to season, Hess argued.

"How animals are maintained prior to being imprinted, between imprint-ing and the test period," had a great impact on imprinting as well.[44]

In the early and mid-1960s, Robert Hinde and Patrick Bateson criti-cally reviewed the research on imprinting, and both concluded that many of Lorenz's ideas had not stood the test of time. Hinde summarized the main points against Lorenz's view of imprinting: "The critical period, ir-reversibility and the extent to which following response affects later be-havior have been overestimated." Bateson thoroughly examined a variety of studies and concluded that Lorenz was right about the effects of im-printing in some bird species, but imprinting was not the special "all-or-nothing" affair uncontaminated by learning that Lorenz had postulated.[45]

These were major challenges to Lorenz's claims about the significance of imprinting and its long-term effects. The idea that there is a critical period during which certain behaviors develop and the claim that the de-railment of imprinting has long-lasting and irreversible consequences for an individual's social relations were the two fundamental tenets of Lo-renz's views about imprinting that psychoanalysts like Bowlby borrowed from animal research. They provided the foundation for claiming a bio-logical basis for similar processes among humans. By the mid-1960s, how-ever, it seemed that those ideas were not even useful for studying birds, let alone other animals. Yet Lorenz did not retreat.

Lorenz's Defense

In 1965 Lorenz published a slim volume, *Evolution and Modification of Behavior*, a revised English edition of a 1961 paper he had published in German defending his original goals for ethology. As he saw it, he now had to contend not only with the challenge from comparative psycholo-gists in the United States, but also with what he called "the English speak-ing ethologists," the defectors Tinbergen and Hinde.

Lorenz began with a defense of the concept of the innate.[46] He distin-guished three attitudes toward the innate. The "majority of American psy-chologists," whom he called the "Behaviorists," held that the division of behavior into innate and learned components was not "analytically valid." According to Lorenz, they presented two arguments to support this the-sis. He considered Donald Hebb (who was neither American nor a behav-iorist), the main defender of the first argument, namely, that the innate was identified only by exclusion and could not be considered a unitary category identifying a specific neural mechanism or process. Lorenz attributed to Lehrman "the Second Behavioristic Argument," the idea that the concept of innate behavior is without heuristic value "because it will never be practically possible to exclude the participation of learning in the early ontogenetic process in the egg or *in utero*, which are inaccessible to observation." Lorenz quickly dismissed this argument by saying it overestimated what could be learned in utero. After three pages dealing with Lehrman's extensive critique by focusing on only one example Lehrman had borrowed from another researcher, Lorenz ratified his steadfast belief in the existence of innate behavior.[47]

Tinbergen and "a great number of other modern ethologists" represented a third position, claiming that all behavior owes its adaptedness both to the phylogenetic process of adaptation and to the adaptive modification of behavior during an individual's life. Thus, Lorenz noted, behaviors formerly described as "innate and learned," respectively, now were taken to "represent only the two extremes in a continuum of gradation in which all possible mixtures and blendings of the two sources of adaptation can be found." For Lorenz, the members of this third group believed that it is not possible to separate, in principle or in practice, the two types of factors influencing behavior. Lorenz characterized this position as "highly dangerous."[48]

More generally, Lorenz portrayed all of his critics' understandings of behavior as "dangerous" and ignorant. They showed "a very deep misunderstanding of biological ways of thinking," while his own views followed as "a matter of course" to "anyone tolerably versed in biological thought." It is not clear who was passing the test of competence in biology, since he was attacking both the major US and Canadian comparative psychologists and the English ethologists, including Tinbergen and Hinde.[49]

The thrust of Lorenz's argument was that his opponents misunderstood the significance of the adaptedness of organisms and the need to separate the two mechanisms that could account for it, phylogenesis and individual learning. This, he believed, was fundamental to the science of ethology: "Ethology loses its character of a biological science if the fact is forgotten that adaptedness exists and needs an explanation." The ethologist was concerned not with individual learning but with phylogenetic adaptations. To investigate the latter, he agreed with his opponents that one should look at the ontogeny of behavior. Yet, he claimed, "any investigator of the ontogeny of animal behavior is well advised to begin with the time-honored procedure of first searching for whatever may be blueprinted by heredity."[50] For that purpose, Lorenz also reasserted the value of deprivation experiments.

Deprivation or isolation experiments, Lorenz argued, would prove that what the environment is doing is "decoding." That is, the environment merely releases what is already encoded in the genome: "What rules ontogeny, in bodily as well as in behavioral development, is obviously the hereditary blueprint contained in the genome and not the environmental circumstances indispensable to its realization."[51]

In this response to his critics, Lorenz now adopted terms from modern genetics, such as genome, blueprint, and decoding, to support his earlier view that social behavior in animals is determined genetically or, as he put it now, "coded" in the genome. Though he ostensibly updated his views by packaging them in the new vocabulary of genetics and molecular biology, he did not modify his research program or his conception of instinct. Nor did he present additional experimental or observational evidence to meet his critics' objections. Instead, he claimed they misunderstood his work because they did not know basic biology.

Lehrman Redux

Before Lorenz's 1965 publication, Lehrman had been optimistic about the increase in common ground shared with the ethologists. He participated regularly in ethological meetings and continued his personal and professional exchanges with English ethologists, especially Hinde and Tinbergen. In 1963 he asserted that one should not overestimate the differences between "ethologists" and "comparative psychologists."[52] Thus the 1965 publication of Lorenz's attack came as a blow.

By this time Lehrman was a world leader in studies of hormones and behavior. He directed the Institute of Animal Behavior at Rutgers University in Newark, New Jersey, a pioneering laboratory in the study of the neuroendocrine basis of behavior. As a specialist in studies of reproductive and parental behavior, Lehrman had been instrumental in showing how behavior can influence hormonal changes. There is feedback from behavior to the endocrine system, and external stimuli (including those from another animal) affect hormone secretion. Lehrman's work was widely respected in the scientific community.[53] Now Lehrman answered Lorenz, fittingly, in a tribute volume honoring Schneirla, who had died unexpectedly in 1968.

Noting the importance of labels as weapons, Lehrman asked why Lorenz called the criticisms that he, Hebb, and Schneirla had raised "behavioristic arguments." The term behavioristic, observed Lehrman, was an affront to the memory of Schneirla, whose work was not in the tradition of American "rat psychology." By labeling his critics in this way, Lorenz was questioning their knowledge of biology, a position Lehrman found prejudiced and disrespectful. He now regretted the elements of hostility in his 1953 critique of Lorenz's theory and expressed dismay at the antagonism in Lorenz's recent defense.[54] But Lehrman stood by his earlier criticisms and did not think Lorenz had addressed them successfully.

First, Lehrman argued that Lorenz's use of the word innate conflated two meanings of "inherited"—heritability and developmental fixity. That a characteristic is heritable does not mean it is not influenced by the environment, he noted.

In trying to assess whether a characteristic is heritable and whether it is influenced by the environment, one is posing two different questions. One question is about the genesis of a characteristic, the other is about its development. As Lehrman put it: "That a characteristic of an animal is genetically determined in the sense that it has been arrived at through the operation of natural selection does not settle any questions at all about the developmental processes by which the phenotypic characteristic is achieved during ontogeny."[55] Clearly annoyed by Lorenz's reiterated accusation that those who did not agree with him did not understand biology, Lehrman retorted:

> If a scientist is not overwhelmingly convinced that characteristics incorporated into the species by the actions of natural selection are, BY THAT FACT, demonstrated to be impervious to individual experience, he is not necessarily guilty of "a very deep misunderstanding of biological ways of thinking" or "a lack of acquaintance with phylogenetic and genetic thought." (Lorenz, 1965)[56]

Lehrman also argued that saying a behavior pattern is innate when it is blueprinted in the genome does not explain much. To say a behavior is innate if it is "blueprinted in the genome" or "encoded in the DNA" may be "comforting, in the sense that it gives us the feeling that we have increased our understanding of the problem," but he emphasized how the use of an analogy can obscure rather than reveal differences.[57]

Lehrman faulted Lorenz's analogy of the relationship between a blueprint and the structure it represents to the relationship between the genome at the zygote stage and the phenotype of an adult organism. As Lehrman pointed out, only the blueprint is isomorphic with the structure it represents. The ratios of lengths and widths in the blueprint are the same as in the structure; topographical relationships among the parts are also the same; each part of the structure is represented by a corresponding part of the blueprint; and each part of the blueprint corresponds only to a specific part of the structure. But "it will be immediately obvious that this is profoundly different from the relationship between the genome and the phenotype of a higher animal." The characteristics and parts of the phenotype, Lehrman continued, are not "represented" in a gene or set of genes; a gene does not influence only, or even primarily, a given part or characteristic of the organism.[58]

Lehrman correctly pointed out that, despite the enormous advances in recent genetic research, it was still true that "the problems of ontogeny and differentiation of structures in complex organisms . . . have hardly been touched as yet by the recent massive advances in molecular biology. A facile description of the genome as a blueprint gives a misleading impression of understanding a problem that is regarded by modern geneticists as one of the major unsolved problems of biology."[59]

As a result, Lehrman did not see Lorenz's update in his vocabulary as progress. The language of molecular biology provided new rhetoric, but this did not address the conceptual and empirical questions Lehrman had raised in 1953 regarding Lorenz's goal of identifying instinctive behaviors. By this time, as this chapter has shown, many animal researchers had realized the sterility and futility of this goal.

Nobody knew this better than Tinbergen, as he stated in a letter to American psychologist Leonard Carmichael:

> We have all traveled a long way since, say, thirty years ago, when we "ethologists" reacted vigorously against what we considered an overrating of the environmental control of behaviour development. Now, curiously enough, we find ourselves increasingly involved in studying environmental influences, and the switch from the dichotomy between "innate" and "learnt" behaviour (still adhered to, later to my dismay, by Konrad Lorenz and many German workers, in spite of the obvious sterility of this attitude) towards the study of the developmental process has now been made by most of the Anglo-American ethologists.[60]

Lorenz kept reasserting his old views, but Tinbergen was no longer following Lorenz. Neither were his students, or Lehrman's students. Tinbergen at Oxford and Lehrman at Rutgers trained a large proportion of the next generation of students of animal behavior. They, like their mentors, moved beyond Lorenz's conception of instincts.

Some of those students have argued that by the 1960s most animal researchers had abandoned Lorenz's ethological research program. For example, Patrick Bateson and Peter Klopfer write, "We take the view that ethology as a coherent body of theory ceased to exist in the 1950s.... One by one the concepts and theories succumbed to critical analysis and, by the beginning of the 1960s, any vestiges of common belief in an ethological theory of behavior had disappeared." Another important ethologist, Colin Beer, says: "The second half of the 1950s and early 1960s was a period of questioning and dismantling as far as ethological instinct theory was concerned."[61]

Conclusion

During the 1950s and 1960s, researchers in Europe and the United States showed the difficulties inherent in the original ethological goal of studying animal behavior by first identifying an animal's instincts. Here I have reviewed some of their major objections.

Lehrman challenged Lorenz's view of social behavior as determined by instincts and called for new standards of evidence to explain why animals act in certain ways and not others. He emphasized the difficulty of cutting animal behavior neatly into separate pieces. He did not believe isolation experiments could identify components of behavior that are innate, in the sense of being the result of inheritance and not affected by environmental factors. To discover the role of inheritance and environment in a given behavior, a scientist has to investigate its ontogeny. This requires physiological and developmental studies of a specific behavior in a specific species. Further, when that work is done, Lehrman argued, it becomes evident that in most cases behavior and emotions are not simply released automatically by internal or external cues. Indeed, they are the result of complex interactions among experience, chemical and biological factors, and environmental influences. Behavior and emotions, in sum, are not released by instincts; instead they develop as a result of the organism's history. In this picture, maternal behavior is not simply "released" by the round face of a smiling baby, as Lorenz and Bowlby had claimed.

Although Lorenz maintained that his theories of animal behavior had implications for human conduct and social policies, Lehrman disagreed with this as well. He did not think there was any scientific basis for Lorenz's extrapolations from birds to humans.

Many other researchers, including leading figures such as Hebb, Beach, and Tinbergen, eventually agreed with Lehrman's main objections to Lorenz's program. They also questioned the ability of isolation experiments to identify innate behavior and the value of Lorenz's conception of instinct.

Other animal experts challenged important parts of Lorenz's model. Hinde showed the lacunae in the concept of drive and psychohydraulic models of energy. Students of imprinting revealed that Lorenz's views about it turned out to be incorrect: in most species, there are no critical periods in which the filial relationship irreversibly determines an animal's future behavior.

In his defense, Lorenz went on the offensive, accusing his opponents of not understanding "biological ways of thinking." Yet repackaging his views in the language of modern genetics and appealing to his own knowledge of evolutionary biology did not address his critics' specific empirical and conceptual challenges to his theories.

CHAPTER FIVE

Psychoanalysts against Biological Reductionism

Introduction

In talking about maternal instincts and infants' instinctual needs, Bowlby substituted the ethological concept of instinct for the psychoanalytic one. When he presented his views on attachment theory in his 1958 paper "The Nature of the Child's Tie to His Mother," he made it clear that he was using the word instinct in the ethological sense of innate behavior patterns.[1] But could psychoanalytic drives be reformulated as biological instincts, as Bowlby proposed? What did psychoanalysts think of Bowlby's attempt to unify psychoanalysis and ethology? This chapter examines how psychoanalysts responded to Bowlby's attempt to synthesize ethology and psychoanalysis.

I focus on the response of three major Freudian analysts: Anna Freud, Max Schur, and René Spitz. I show that the gulf between them and Bowlby was unbridgeable, both in what they thought counted as facts and in their theoretical interpretations. Robert Karen argues that psychoanalysts Anna Freud, Schur, and Spitz launched a surprise "Freudian assault" to uphold Freudian orthodoxy. Further, in Karen's view many psychoanalysts, including Melanie Klein, rejected Bowlby's position because they were not interested in the facts he had presented about children. This interpretation closely follows Bowlby's own version of this history. According to him, many psychoanalysts rejected his position because they did not know biology and they were not concerned with real external events.[2] Indeed, many psychoanalysts rejected Bowlby's synthesis of ethology and psychoanalysis, but not for the reasons Karen and Bowlby noted—lack of knowledge of biology and lack of interest in children's behavior. In addition, Bowlby was not taken by surprise, for he was deeply involved in the discussions about his work over the years.

To understand these psychoanalysts' reaction to Bowlby's work, I situate it within a larger framework. After discussing Sigmund Freud's views about instincts, I then note some key resemblances between the psychoanalytic and ethological models of the human mind. As we saw in chapter 2, and as I highlight here, Bowlby's work arrived at a moment of unprecedented interaction between students of animal behavior and psychoanalysts. His work encouraged further discussion of the relation between the two fields. Ultimately, however, Freudian analysts Anna Freud, Schur, and Spitz did not accept Bowlby's proposal because they recognized that doing so would amount to rejecting core organizing principles in psychoanalysis.

Freud on Instincts

For Freud, instincts are both the foundation and the engine of the human mind.[3] Yet throughout his life he was aware that instincts were the dark continent of psychoanalysis. Deeply influenced by evolutionary thinkers like Darwin and Lamarck, Freud erected his view of the psyche on a biological foundation, but he also made it clear that psychology is not reducible to biology.

According to Freud, humans are born with a ready-made set of instinctual needs.[4] All individuals experience conflicting impulses during their mental development. Those impulses motivate them, frustrate them, and keep them in a state of constant psychic struggle. In his early writings Freud referred to those impulses as "excitations," "affective ideas," and "wishful impulses." He separated the external excitations that usually affect the mind with a single impact from the internal ones that exert constant pressure. Freud called those internal stimuli instincts (*Triebe*) or instinctual impulses (*Triebregungen*) and said that they are the primal motivating forces of the mental apparatus.[5]

Freud proposed that instincts are housed in the *id*, one of the three parts of the psychic apparatus. According to him, the id contains "everything that is inherited, that is, present at birth, that is laid down in the constitution—above all, therefore, the instincts, which originate from the somatic organization and which find a first psychical expression here (in the id) in forms unknown to us." The second part, the *ego*, is intermediary between the id and the external world and controls the instinctual impulses, "deciding whether they are to be allowed satisfaction." The third part, the *superego*, is the repository of norms and values that individuals accumulate mainly through parental influence.[6] As is clear in this model, the instincts contained in the id are the oldest and most fundamental strata in the mind.

For Freud, an instinct has several parts: source, pressure, aim, and object. The source is the somatic process in the body whose stimulus is represented in mental life by an instinct, but he was unsure whether the source had a chemical nature or involved the release of mechanical forces. The pressure refers to the amount of force the instinct exerts on the mind. The aim is the instinct's satisfaction, which can be obtained only by removing the cause of stimulation at the instinct's source. The object concerns the thing through which the instinct can achieve its aim.[7] This system works pretty much like a hydraulic model, in which energy accumulates until it is released.

For Freud, the first goal was to find the primary instincts operating in all human beings. In 1915 he proposed the ego, or self-preservative instincts, and the sexual instincts as primal.[8] After World War I he introduced a new force, the death instinct, and postulated a duality between two major primal instincts, Eros and Thanatos, the life instinct and the death instinct.[9]

As to the nature of instincts, Freud hoped biology would one day help psychoanalysis. He pointed out that from a "biological point of view, an instinct appears to us as a concept on the frontier between the mental and the somatic, as the psychical representative of the stimuli originating from within the organism and reaching the mind, as a measure of the demand made upon the mind for work in consequence of its connection with the body." He thought it "doubtful" that "any decisive pointers for the differentiation and classification of the instincts can be arrived at on the basis of working over the psychological materials." So "it would be desirable if those assumptions could be taken from some other branch of knowledge and carried over to psychology." Specifically, he hoped that biology would one day clarify the nature of instincts. However, he added that the biological aspects of instincts lie outside psychoanalysis: "The study of the sources of instincts lies wholly outside the scope of psychology. Although instincts are wholly determined by their origin in a somatic source, in mental life we know them only by their aims."[10]

For some authors this separation between the biological and the psychological sides of instincts was muddled by the translation of Freud's writings from German to English. Freud usually referred to the human instincts as *Triebe* and to animal instincts as *Instinkte*. In English, Freud's *Triebe* has normally been translated "instincts." For many psychoanalysts and historical interpreters, this translation is the source of much confusion about Freud's views, since it suggests that he thought human instinctual impulses are the same as animal instincts. These commentators believe translating *Trieb* as "drive" would solve the problem. But I do not think the matter is that simple, because in both English and German "drive" is also used for animal instincts. The important point is to understand Freud's conception of instincts. This is not easy, because he was not always clear about his views and also changed them over time. In several places he recognized that instincts were one of the weakest areas of psychoanalysis. But on one point Freud was clear: psychoanalysts should focus on the mental aspects of instincts.[11]

Freud's conception of instincts as mental phenomena, and not simply biological entities or processes, was fundamental for establishing psychoanalysis as an independent science. It is precisely because there is more to an instinct than biology that psychoanalysis has something to say about the human mind. Establishing the difference between the biological and mental aspects of human instincts led to a separation of the tasks of biology and psychoanalysis. Psychoanalysis would focus on the psychic expression of the instincts and would illuminate the ways human beings deal with the conflicts created by instinctual internal pressures and other external pressures impinging on their minds.

How much Freud wanted to separate biology from psychology is nowhere more evident than in his views about motherhood. For Freud, a woman develops her mature psychological self only through motherhood. Both boys and girls go through the two earliest phases of development, oral and anal, but they part ways when boys confront the Oedipus complex. At this point girls develop penis envy and feel psychologically castrated. For Freud, a girl resolves her inner conflicts only when she gives up her "wish for a penis and puts in place of it a wish for a child."[12] Furthermore, a woman reaches complete happiness only when she gives birth to a son: "A mother is only brought unlimited satisfaction by her relation to a son; this is altogether the most perfect, the most free from ambivalence of all human relationships."[13] Thus, for Freud motherhood plays a central role in women's mental and emotional development.[14] Yet Freud did not posit a maternal instinct. For Freud, a woman's desire for a child is not a primary biological drive; rather, it is a woman's way of compensating for her lack of a male body.

Not all psychoanalysts agreed with him on this point. In 1926 one of his followers, Karen Horney, asked "in amazement" about the maternal instinct. She argued that it was a primary urge, "instinctually anchored deeply in the biological sphere."[15] Freud, however, criticized the increasingly deterministic vision of biology that he thought Horney and other psychoanalysts were defending. In private correspondence, he left no ambiguity about this. In a 1935 letter he wrote: "I object to all of you to the extent that you do not distinguish more clearly and cleanly between what is psychic and what is biological, that you try to establish a neat parallelism between the two. . . . [W]e must keep psychoanalysis separate from biology just as we have kept it separate from anatomy and physiology."[16]

The nature of the psychoanalytic "drives" and the possible existence of a maternal instinct would remain central points of disagreement among psychoanalysts. Some of the most prominent early followers of Freud— such as Karen Horney, Helene Deutsch, and Alice Balint—debated whether women's motherliness and maternal feelings are primary instinctual drives. Other psychoanalysts attempted to clarify the notion of instinct/drive in psychology.[17]

It is in this context of debate over the nature of instincts and their role in shaping the human psyche that we need to understand how psychoanalysts reacted to Bowlby's attempt to unify ethology and psychoanalysis. Freud's definition of instinct as a concept between the mental and the somatic and his hope that biology would clarify the nature of instincts also contributed to the appeal ethology held for Bowlby. It is therefore worth reflecting on the similarities between the psychoanalytic and ethological models of motivation.

Psychoanalysis and Ethology: Natural Allies?

"Freud stands squarely within an intellectual lineage where he is, at once, a principal scientific heir of Charles Darwin and other evolutionary thinkers in the nineteenth century and a major forerunner of the ethologists and sociobiologists of the twentieth century." So asserts Frank Sulloway, one of Freud's major biographers.[18] To call Freud a forerunner of ethology without further analysis seems an overstatement, but ethology and psychoanalysis share certain basic ontological and epistemological tenets.

It is difficult to assess Freud's impact on Lorenz. As we saw in chapter 2, after Germany annexed Austria in 1938, Lorenz referred to Austrian psychology as derived "from the thought of Jewish-babbling, verbose, Jewish leaders."[19] I have not found Lorenz referring to psychoanalysis until the postwar period. Both the absence of earlier citations and the references later can probably be explained by opportunism, or at least partly so. It was not politically advantageous to cite a Jewish thinker during Nazi times, but Freud's thought underwent an important revival after World War II. As I argue in several parts of this book, Lorenz's work benefited from the interest that psychoanalysis had created about instincts and the effects of early child rearing on a child's emotional makeup. Despite the lack of direct evidence, we can safely assume that Lorenz was familiar with Freud's work. Like Freud, he grew up in Vienna, trained in medicine, and aimed to situate psychology on a foundation of natural science. In addition, his mentor in psychology, Karl Bühler, had examined psychoanalysis in his writings. Although later in life Lorenz denied having been influenced by psychoanalysis, several contemporaries noted the resemblance of some of his ideas to psychoanalytic views. Indeed, some components of the ethological framework are similar to the psychoanalytic one.[20]

Much like Freud, who talked about instincts as innate needs, Lorenz said organisms have innate appetites and moods established by the phylogenetic history of their species. Furthermore, Lorenz's conception of instincts is very similar to Freud's. As I discussed in chapters 2 and 4, Lorenz put forward a psychohydraulic model, where instincts work though internal mechanisms that control energy accumulation and discharge. In the same way that Freud spoke of instincts as endogenous or internal stimuli that put pressure on the mind, Lorenz claimed that instinctual energy builds up inside the organism, creating pressure to discharge.

In explaining behavior, both Freud and Lorenz emphasized the existence of motivational conflicts. The conflicting pressures in psychoanalysis lead to anxiety. In the case of animals, Lorenz talked about the "restlessness" produced by accumulated energy. In both models, the release of the energy leads to "satisfaction," and unused pent-up energy is eventually released onto "inappropriate" objects. Animals and humans engage in displacement. Humans can also sublimate their drives.

Another common postulate in psychoanalysis and ethology is the existence of critical periods in early infancy that determine the pattern of social relations in adulthood. Instincts are central here as well, for they guide the unfolding of individual development along a natural path, which conforms to the typical, universal path for normal development. If anything goes wrong during these early developmental stages, it can cause permanent damage later in life. Animals, like Lorenz's ducks, could also develop "fixations," or close attachment to inappropriate objects, as happened when birds imprinted on him.

Furthermore, in both frameworks, the relationship with the mother is of special importance for the individual's normal development. The mother-infant relationship determines all the individual's future social relationships, sexual responses, and love interests. According to the psychoanalytic and ethological frameworks (at least, as many people interpreted Lorenz's work on imprinting), the individual has to develop a bond with the mother, and then separate from her to reach maturity, in order to establish normal relationships with the members of the same species.

Given the many similarities outlined above, it is not surprising that Bowlby believed ethology could clarify important issues in psychoanalysis. As we saw in chapter 3, he appealed to Freud's authority to justify his turn to biology, since in several places Freud had expressed his desire that biology would one day help psychoanalysis illuminate the nature of instincts. In fact, Bowlby was not the only psychoanalyst interested in ethology.

During the late 1950s and early 1960s, students of animal behavior and psychoanalysts engaged in an intense exchange of ideas. Many psychoanalysts were interested in ethology and receptive to Lorenz's work. After a trip to the United States in 1958, Lorenz reported: "I am just back from America where I have been talk-talk, talking to Psychiatrists, Psychoanalysts and Psychotherapists."[21] Some psychoanalysts attended conferences on ethology and even visited ethological research centers. In 1959 René Spitz spent a few months at Lorenz's experimental station at the Max Planck Institute in Seewiesen, Germany. Many psychoanalysts had studied biology, especially those in the United States, where psychoanalysts were required to have medical training.

Immediately after Bowlby's presentation of his theories, an intense discussion of ethological ideas took place within the psychoanalytical community. As expected, many discussions concerned the concept of instinct in ethology and Bowlby's importing this concept into psychoanalysis.[22]

Despite their interest in biology, most psychoanalysts rejected Bowlby's synthesis of ethology and psychoanalysis. As Bowlby later recalled it, the first major confrontation took place in England with the followers of Melanie Klein, after Bowlby presented his views at the Tavistock Institute. Bowlby maintained that this rejection was due to Klein's dismissal of actual experiences as a key to understanding trauma. For her, to understand childhood disorders it was fundamental to pay attention to fantasy, especially to the child's construction of the image of the mother. Some Kleinians may have been extreme in their rejection of real-life experiences and relations with parents, but many psychoanalysts criticized Bowlby's position because he focused *only* on real-life experiences and thus ignored the internal psychological processes that are the focus of psychoanalysis. Even psychoanalysts who were not at all in Klein's camp rejected Bowlby's proposal.

Here I will examine in greater detail the reactions of Anna Freud, Max Schur, and René Spitz. I focus on them because of their high status in the psychoanalytic community and their influence in the United States; because they presented the first major criticism that appeared in print; and because Bowlby claimed their objections "acted as big red lights warning less senior analysts" to reject his views.[23]

Anna Freud

As we saw in chapter 1, Anna Freud had been one of the earliest analysts to call attention to the significance of the child's environment and the mother's role. From the start, she was also interested in Bowlby's work. They kept in contact through frequent friendly correspondence and occasional visits to each other's units. Given Anna Freud's interests, experience, and status in the field, Bowlby wanted her as an ally. It is clear from their letters that he courted her approval and support.

When Bowlby took up his appointment with the World Health Organization, he wrote to Anna Freud about his study, asking her about relevant psychoanalytic literature on the mother-child relationship. He also told her he wanted to do a follow-up study of the children from the Hampstead Nurseries with James Robertson, who had worked with Anna Freud before moving in 1948 to the Tavistock Clinic to work with Bowlby.[24]

Robertson remained in close contact with Anna Freud, thus connecting her unit and Bowlby's. Robertson still helped Anna Freud with

FIGURE 5.1. James Robertson with Anna Freud at 20 Maresfield Gardens, her home in London, on the occasion of the Freud centenary celebration in 1956. Freud Museum, London.

Aftercare, a project designed to monitor children who had been at the Hampstead Nurseries when it was run by Anna Freud and Dorothy Burlingham during the war, and who were now in foster care. The Hampstead Clinic kept in touch with many of the children and wrote annual reports on their progress and current situation for their funding agencies. These follow-ups included an annual visit by Anna Freud and Robertson around Christmas. Robertson continued to be in charge of writing many of the reports and buying some presents for the children and their families. Bowlby notified Anna Freud that he and Robertson wanted to assess the present adjustment of thirty of these children, but they later abandoned the project because finding enough information became too difficult.[25]

Bowlby told Anna Freud about his work in considerable detail. After his WHO assignment, he presented his results in two talks at the Tavistock Institute during June 1950. He sent an invitation to Anna Freud and her staff.[26] He also wrote that month asking permission to quote from her writings.[27] Bowlby obviously cared about her views and sought her approval before publication. When he finished the WHO report, he wrote to her again and sent her a copy.[28]

In addition, at this stage her support for Robertson was to a great extent also support for Bowlby, since the two men were planning several joint publications. Robertson was filming how children who were hospitalized reacted to being separated from their mothers. Through the showing and discussion of Robertson's film, Anna Freud was also informed about the work done at Bowlby's unit. After some discussion about dates, on November 16, 1951, Robertson wrote to let Anna Freud know that he and Bowlby would show his first film to the Hampstead Child Therapy Course on December 6, 1951. Later they planned to visit again, to continue discussion of the film on January 24, 1952.[29] A couple of years later, when Robertson went to the United States to show his film, Anna Freud wrote to several important psychologists and psychoanalysts there to introduce him and show her support for his work.[30] This helped Bowlby as well, because Robertson's silent film had an accompanying text that referred to Robertson's and Bowlby's publications and joint work.[31] During this period, Bowlby and Robertson published several papers together. Not only did they present Robertson's observational data, but they also started to develop an explanatory framework for the empirical observations. In a 1952 article they proposed that children separated from their mothers go through three phases: protest, despair, and denial (later protest, despair, and detachment). With psychologist Mary Ainsworth, who was also at Bowlby's unit at the time, Robertson and Bowlby were planning a joint book on the effects of maternal separation.[32]

In the early 1950s, then, Anna Freud and Bowlby had a friendly relationship. She supported his work. In turn, Robertson and Bowlby continued to use her studies from the Hampstead Nurseries to elaborate their views on children's emotional development.

From 1953 to 1958, however, there is no correspondence between

Anna Freud and Bowlby. Although in part this could be due to contingent factors unrelated to their relationship, it also indicates increasing differences between them. As we saw in earlier chapters, soon after publishing his WHO report, Bowlby had encountered the work of Konrad Lorenz, and from 1953 on he met with ethologists and comparative psychologists on numerous occasions. But while Bowlby was looking to biology for confirmation of his views about the tremendous power of mother love, Anna Freud was moving in the opposite direction.

In 1954 Anna Freud objected to the increasing blame put on mothers for their children's pathological conditions. She noted that the idea of maternal "rejection" had become so vague and overused as to be "almost meaningless." She argued "against the error of confusing the inevitably frustrating aspects of extrauterine life with the rejecting actions or attitudes of the individual mothers." In addition, she did not think her work with children during World War II provided evidence for a maternal instinct. After separating from their children during the war, many mothers felt detached from them. In her view, "such experiences make one feel that the mother's attachment is bound up with the appeal made by an infant's helplessness and urgent need for care rather than being a mere 'instinct.' "[33]

The issue of how much children were affected when separated from their mothers created tension between Anna Freud and Bowlby's group. Following a public discussion of this point, Robertson urged Anna Freud to realize how difficult it was to clarify things in a large meeting. As for Bowlby's view, he noted, "All I wanted to say on the first point was that Dr. Bowlby is now very aware that although early separation (particularly if accompanied by prolonged deprivation) is likely to lead to damage, the outcomes are, in fact, much more varied than he had stated in his earlier works."[34]

Another indication of building tension over Bowlby's position comes from a letter sent to him by another psychoanalyst. Donald Winnicott and Bowlby had a long-standing personal relationship, and their views were very similar in many respects. In the psychoanalytical world of post–World War II England, they staked out a middle ground between the supporters of Melanie Klein and of Anna Freud. Although Winnicott did not criticize Bowlby in print, he did express concern about the practical uses of Bowlby's ideas. In 1954 he wrote to Bowlby that his work was being used "by those who want to close down Day Nurseries." He urged Bowlby to make a strong public statement that such measures were not desirable, since in the absence of day nurseries, "mothers needing help put their children out with unqualified and unregistered foster parents."[35] I have not found a reply from Bowlby, but we know he did not follow Winnicott's advice.

In the mid-1950s, Anna Freud believed that Bowlby's views about maternal deprivation and separation went beyond the evidence from the Hampstead Nurseries and the data Robertson collected from hospitals. Here we see a budding disagreement about the extent of mothers' responsibility for their children's problems. Yet it seems safe to assume that no animosity had developed between them, since the correspondence picks up again a few years later in a cordial and respectful tone.

Bowlby's attempt to synthesize psychoanalysis and ethology in the late 1950s widened the rift between him and Anna Freud. On October 27, 1958, Bowlby asked Anna Freud to consider his manuscript "Separation Anxiety" for the *Psychoanalytic Study of the Child*. He noted that he was going to present part of this paper at the British Psychoanalytical Society's meeting of November 5 and that this would be part of a forthcoming book with Robertson.[36] In her reply, Freud said she had no decisive voice in the journal but would forward his paper to the editor, psychoanalyst Ruth S. Eissler.[37] Eissler offered to publish it the following year, and from Bowlby's response we can gather that she proposed to present it together with some comments. Bowlby decided to submit it to another journal that could bring it out sooner.

He also suggested submitting instead another paper, "Grief and Mourning in Infancy," that he would have ready in the fall, adding that he would welcome having some comments printed along with his contribution:

> Since like SEPARATION ANXIETY it is controversial (though I hope the style is not polemical), the proposal to publish comments in the same number is very welcome. It is already in advanced draft and I hope to read it to the British Psychoanalytical Society in the Autumn. There should therefore be plenty of time for Miss Freud and others to prepare their contributions before Volume XV goes to press.[38]

The editor agreed, and the paper appeared with critical commentaries by Anna Freud, René Spitz, and Max Schur.[39] Historians of attachment theory have presented the publication of the commentaries as representing the official repudiation of Bowlby's work by the gatekeepers and have suggested that he was overwhelmed by a surprise attack.[40] However, as the correspondence makes clear, Bowlby was aware of their concerns and instrumental in the 1960 publication of the responses along with his paper. Perhaps he was surprised to find how critical the commentaries were, but even this seems unlikely, because he knew the commentators' views through private and public discussions.

Bowlby read Anna Freud's comments before publication, wrote to her, and, in light or her comments, made some corrections to his paper before submitting it.[41] Bowlby even sent Anna Freud his revisions to make sure he had presented her views "accurately and fairly."[42] Furthermore, before she put her opinion in print, they had discussed their views several times, and their differences had become public on several occasions.

Anna Freud took issue with Bowlby's use of her work to support his ideas about the mother-infant dyad. On November 5, 1958, Bowlby read an early version of his 1960 paper "Separation Anxiety" before the British Psychoanalytical Society. In this paper Bowlby presented the idea that children pass through three stages when separated from their mothers: protest, despair, and denial.[43] Anna Freud noted that she felt compelled to comment on Bowlby's paper because he referred to her writings with Burlingham. She divided her comments into two sections, "Agreement about Observed Facts" and "Disagreement about Interpretation."[44]

To refresh the attendees' memories, she reported that the separation theories based on the work done at the Hampstead Nurseries derived from "close acquaintance" with about eighty separated infants and young children, ranging in age from ten days to five years, over a period of four years on a "twenty-four-hour basis." Several caretakers carried notebooks and wrote down their observations about the children. She listed the factors that were taken into account: communications by the children, expressions of their affects and moods, manifestations of libidinal and aggressive impulses and fantasies, progressive and regressive moves in their development, and any changes in their health. Then Anna Freud added: "This material was evaluated on the basis of analytic knowledge, namely, by comparison with the mental states seen or reconstructed in analysis."[45] This last note is fundamental because it made it clear that their evaluations of children did not rest only on observation of their behavior.

Based on that work, Anna Freud agreed with Bowlby about the "observed facts." When children were separated from their families, there was a first stage of protest, longing, and hope for their prompt return. This gave way to a second stage of anger and despair. Later, children entered a third stage of withdrawal, which often included severe regressions in their development. In his presentation, Bowlby had asked why Anna Freud and Burlingham did not call the whole process "separation anxiety," as he referred to it. She responded that they observed anxiety in children not on separation from the mother but on her return, when a child feared she might leave again.[46]

The disagreement over the significance and interpretation of those facts was deep. Bowlby relied on ethology and assumed that the primary tie to the mother was biological. But Anna Freud emphasized that the concept of anaclitic object choice postulated by psychoanalysis was a psychological assertion. As such, she continued, it "deals with the impact on the mind made by the acts of mothering, namely, with the pleasure-pain experiences which accompany primary instinctual reactions and form their mental content."[47]

Thus, Bowlby could not have been surprised by Anna Freud's comments on his 1960 paper "Grief and Mourning in Infancy" for the *Psychoanalytic Study of the Child*. In fact, her response simply elaborated on her earlier points. In her paper's first section, "Identity of Observations," she stated that there was "little difference in the observed material collected during the war by the Hampstead Nursery team with regard to separated children and the observations made later in connection with Dr. Bowlby's study of separation anxiety by a Tavistock Clinic team with regard to hospitalized children." She pointed out that Bowlby was mainly relying on observational work carried out by James Robertson, who was a "valued member" of both the Tavistock team and the Hampstead group. This "identity of material, and partly of observers," made it especially important to explain the divergence in their interpretation and, as she put it, "why misunderstandings in the discussion of the divergences are persistent."[48]

Anna Freud proposed that Bowlby's position could be divided into three parts: "a biological theory in which an inborn urge is assumed that ties an infant to the mother"; "the behavior resulting from this tie ('attachment behavior')" or from its disruption (separation anxiety, grief, and mourning); and the events and actions in the external world that "activate inherited responses" that thus bridge the gap between the biological urge and the manifest affect and behavior. She then noted that between her interpretation and Bowlby's there was, as she put it, a "clash between metapsychological and descriptive thinking." While praising Bowlby's regard for biological and behavioral considerations, she argued that taken by themselves, not in conjunction with metapsychological thinking, these two types of data do not fulfill the analyst's requirements. As analysts we do not deal with drive activity as such but with the mental representation of the drives. . . . Equally, we do not deal with the happenings in the external world as such but with the repercussions in the mind, i.e., with the form in which they are registered by the child.[49]

Anna Freud recognized that translating behavioral data into psychological terms introduced "numberless complications" not present in Bowlby's simpler account. Yet she found this to be necessary if one sought a "true reflection of the complexity of mental life." This complexity included the interplay between drives, the sensations and perceptions from the internal and external world, experiences of pain and pleasure, fantasies, and as the child grew older, mental processes such as verbal and logical thinking and conflicts such as anxieties, guilt reactions, and defensive activities.[50]

She expanded on other differences between her views and Bowlby's as well, to clarify the specific role of the pleasure principle in childhood, infantile narcissism, and the phases of behavior after separation from the mother. Still, the one difference that made their positions irreconcilable stemmed from what she called "the clash between metapsychological and descriptive thinking."[51] More specifically, she noted that Bowlby's scheme eliminated psychological processes and thus the psychoanalytical apparatus needed to understand them.

In addition, Anna Freud criticized Bowlby's use of data on hospitalized children to understand separation between mothers and infants in everyday circumstances. As she noted, those children were often separated from their parents for long periods, and under stressful conditions like war. She even emphasized the limitations of her own studies, stating: "Neither the Hampstead Nurseries nor hospitals and other residential homes have offered ideal conditions for studying the length of time needed by young children to displace attachment from one person to another." Both sets of data concerned children who not only had lost their mothers but had also been separated from other family members and all other objects in their familiar surroundings. In addition, they had to adapt to group life. Given those circumstances, Anna Freud proposed that "from direct observation we know little or nothing about the duration of grief in those instances where the mother has to leave temporarily or permanently while the child remains at home."[52]

Contrary to Bowlby's assertions that there was incontrovertible proof of the detrimental effects of separation from the mother and the negative consequences of day care, Anna Freud admitted that scientists knew "little or nothing" about the effects of those separations. In her opinion Bowlby had based his conclusions on problematic data obtained mostly in hospitals, wards, and other institutions where children were deprived of much more than their mothers.[53]

On January 4, 1961, Bowlby wrote to Anna Freud, thanking her for the "spirit" of her comments. Of her kindness there is little doubt, since she made an effort to find points of agreement. But of the immense gap between their positions, there can be no doubt either.[54]

In fact, even Anna Freud's assertion that she and Bowlby agreed on the data is confusing. It is true that she had been among the first to document how children suffered when separated from their parents during the war. However, in the reports on the Hampstead Nurseries that she published with Dorothy Burlingham, one can read over and over that what caused children's regressions was not separation per se, but *the manner* in which the separation took place. For example, in discussing the attitudes of children toward bombing, they said that "the children's fears are to a larger extent dependent on their parents' anxiety." Burlingham and Freud then documented how the children's fears decreased after separation from their parents. Reporting on the case of Billie, a child who suffered from self-control problems and compulsive behavior that disappeared when his mother returned, they wrote: "The interesting point about this story is that it does not seem to be the fact of separation from the mother to which the child reacts in this abnormal manner; rather, he seems to react primarily to the traumatic way in which this separation took place. Billie can dissociate himself from his mother when he is given three or four weeks to accomplish this task. If he has to do it all in one day, it is a shock to which he responds with the production of symptoms."[55] After reporting on numerous cases, Freud and Burlingham concluded, "Our case material shows that it is not so much the fact of separation to which the child reacts abnormally as the form in which the separation has taken place. The child experiences shock when he is suddenly and without preparation exposed to dangers with which he cannot cope emotionally."[56]

Indeed, there is so much emphasis on the manner of separation, and not on separation per se, that one cannot help wondering why Anna Freud emphasized their agreement on facts in her two discussions about Bowlby's work. Her desire to recognize the value of Robertson's work may have played a role. By separating the theory from the observations, she could ensure that his work on separation in hospitals was not rejected in conjunction with Bowlby's work. This strategy also highlighted the fact that Bowlby did not have his own set of observations to support his theories.

We can conclude that Anna Freud differed from Bowlby on three fundamental points. First, being deprived of the mother per se is not the immediate cause of a child's sense of bereavement. Second, the results about separation obtained in hospitals and war nurseries cannot be extrapolated to ordinary circumstances. And third, psychoanalysts must focus on the mental life of the child; therefore they cannot accept a reduction of mental drives to biological instincts. Given this goal, psychoanalysis could not rely only on behavioral data.

Anna Freud's criticisms of Bowlby were fundamental, and they also stemmed from her work with children. Interpreting them as mere defenses of her father's legacy shortchanges her research and contributions to child analysis.

Max Schur

Max Schur was born in 1897 in Stanislaw, part of the Austro-Hungarian Empire in what is now the Ukraine. After fleeing with his family from the advancing Russian army, he finished his secondary education in Vienna and later obtained an MD from the University of Vienna. While doing medical research and practicing as a doctor, he received psychoanalytic training, and in 1931 he became an associate member of the Vienna Institute. He was Freud's personal physician from 1929 until Freud's death in 1939 and was a close friend of the family. Freud helped arrange his departure from Vienna when his own family left for England. In the early 1940s Schur established himself in New York, where he became a leader in psychosomatic medicine and served as president of the American Psychoanalytic Association. He continued his work as physician and psychoanalyst until his death in 1969.[57]

Anna Freud seemed to value Schur's work in psychoanalysis, his medical advice, and his friendship. In their continual and frequent correspondence, he remained "Dr. Schur" to her, but he was one of the few people who addressed her as "Anna." Schur often sent medical advice and even prescriptions for Anna Freud and Dorothy Burlingham. Anna Freud ex-

FIGURE 5.2. Max Schur with unidentified woman, Topeka, Kansas, 1966. Freud Museum, London.

pressed her high regard for his work, which she considered "the best" in the area of psychosomatic medicine, which was expanding in midcentury America.[58]

When Schur was thinking about his upcoming presentation for the 1959 meeting of the Psychoanalytic Association in Copenhagen, Anna Freud encouraged him to address the "*Triebthema*" (the topic of drives) in biology and in psychology. In her view it was necessary to clarify misunderstandings from previous work in the area, so that "nobody could confuse the biology for the psychology." Agreeing with her advice, he prepared a talk on that topic. Then Schur heard there was to be a panel on ethology and psychoanalysis under Bowlby's direction. As he told Anna Freud, he wrote to the organizer of the conference, Charles W. Tidd, that he would like to present a paper in Bowlby's session.[59] Schur continued to keep Anna Freud informed about the development of his ideas and presentations, including one for another symposium in Chicago, referred to earlier, where he noted the importance of Schneirla's work, which he found more interesting and complex than Lorenz's and Tinbergen's.[60]

By the time Eissler asked Schur to comment on Bowlby's paper for the *Psychoanalytic Study of the Child*, Bowlby and Schur had already engaged in several exchanges about the relation of biology to psychoanalysis. Schur wrote to Anna Freud that he would do the comment for the journal "because she clearly found it necessary," but his experience in Copenhagen and his later correspondence with Bowlby led him to believe it would be a waste of time to try to convince Bowlby. Schur pointed out two major faults with Bowlby's work. The first was that Bowlby focused exclusively on the child's instinct to cling as the vehicle for attaching to the mother, with no role for the need for food. The second was his misunderstanding of the difference between instinctual drive and instinct. Schur developed these points in his published commentaries on Bowlby's work.[61]

In his response to Bowlby's paper on separation anxiety, Schur first noted that the article under discussion built on Bowlby's earlier papers. Thus he would comment on Bowlby's larger effort to provide a synthesis of ethology and psychoanalysis. Like Anna Freud, Schur started by agreeing with Bowlby on the significance of the infant's tie to the mother and the intensity of the child's reaction to separation. But he disagreed with Bowlby's explanations.[62]

Schur mainly objected to Bowlby's understanding of instincts. As he saw it, his theory incorporated that "part of the instinct theory of ethology which assumes the *fully innate, unlearned character of most complex behavior patterns*." However, this theory was not widely accepted in biology and psychology, as indicated by the strong opposition from the "'biopsychologists' represented in the United States ... by Schneirla, Lehrman, Beach, and others." By assuming that human infants "are starting with a number of *highly structuralized responses*," Bowlby was applying to human behavior a concept of instinct that neglected the importance of development and learning.[63]

Furthermore, Schur said that by employing a purely biological concept, Bowlby was backtracking from Freud's advances. For Schur, one of Freud's greatest contributions was his move from a "'biophysiological' to a psychological concept" of instinct. This transition was the basis of any psychoanalytic theory of motivation. In contrast to need satisfaction, which is a physiological concept, psychoanalysts deal with mental wishes and the pleasure-unpleasure principle, which operates at the psychological level. Thus, like Anna Freud, Schur accused Bowlby of eliminating the psychological aspects of instincts. In addition, he questioned the biology Bowlby was using.[64]

Anna Freud told Schur she had asked Ruth Eissler not to "surprise" Bowlby, but to let him know about the orientation of the comments. She mentioned that she had sent her comments to Bowlby and suggested Schur do the same. It is not clear whether he did so. In any event, in the same letter where he thanked Anna Freud for the spirit of her comments, Bowlby mentioned that he regretted that the other commentators, Max Schur and René Spitz, had not "followed suit."[65]

Bowlby was particularly upset about Schur's commentary. In a letter to Anna Freud, he complained that Schur misrepresented his ethological views. Bowlby included a copy of a letter he had written to Max Schur on the same date, and which he said had circulated among several analysts. In that letter he pointed to several places in his own writings where he noted the central role of learning in the child's emotional development. In addition, Bowlby said he found it shocking that Schur could attribute to him support for the hydrodynamic model of energy that Lorenz had postulated. Bowlby pointed out that he had never referred to "action-specific energy" and had explicitly noted in his papers that ethologists had abandoned Lorenz's hydrodynamic theory. He felt that much of Schur's case would stand or fall on his assertion that Bowlby's theorizing was not "well and truly rooted in modern biological thought." To justify his views as not outlandish, Bowlby noted that comparative psychologist Frank Beach had read his paper "The Nature of the Child's Tie to His Mother" in 1957–58 when they were together at the Behavioral Sciences Center at Stanford. He also claimed that Robert Hinde, Niko Tinbergen, and Harry Harlow were his correspondents and in general supported his work.[66]

It was indeed true that as early as 1957 Bowlby had noted that the psychohydraulic model was discredited. (Bowlby mentioned that Lorenz and Tinbergen were no longer using it. However, Lorenz continued to uphold it until the end of his career.) In fact, Bowlby said he wished psychoanalysts would abandon it one day as well.[67] How far animal researchers supported Bowlby's views is more debatable. In any case, the key point under discussion was whether psychoanalysis would gain or lose by adopting the ethological concept of instinct that Bowlby borrowed from Lorenz to explain children's emotional needs.

Bowlby and Schur both kept Anna Freud informed of their opinions of each other and their disputes. Aware of Bowlby's frustration about his commentary, Schur told Anna Freud that he found Bowlby's reaction to the critical commentaries, as expressed in Bowlby's circulated letter, "paranoid." In Schur's view, Bowlby merely gave lip service to the role of learning. Yes, Bowlby said "he believes in learning." But when he presented his views, he ignored learning, just as he ignored the development of the ego. In addition, he continued to present instincts as fully formed mechanisms that are merely "activated."[68]

Paranoid or not, Bowlby was mindful of how much was at stake. His main claim to originality was his development of an ethological paradigm within psychoanalysis. Recognition that his views were grounded in legitimate biology was paramount for the success of his theory. As he put it in the letter to Schur: "Since a main object of my work is to link psychoanalytic theory with modern biological theory, the inaccurate way you have reported my views on it is of some consequence to me." Bowlby noted that he was sending copies of this letter to a number of analysts on both sides of the Atlantic, and he hoped Schur would set the record straight in other publications. For a public audience, Bowlby was drafting a response to Schur along the lines of the letter.[69]

The publication of his response became a saga. He first tried to publish it as a response in the journal where his original paper and the critical commentaries had appeared, but Eissler reported that it was not the journal's standard practice to publish rejoinders. When Bowlby appealed to Anna Freud, she noted that she was not supposed to interfere with editorial policy but would write to the editor anyway.[70]

Schur thought Bowlby was lost to the cause of psychoanalysis, but Anna Freud insisted he could listen to his critics. On February 4, 1961, she wrote to Schur: "My experience with Bowlby shows without a doubt that he reacts very well when one answers all his arguments in an objective and not personal way." Anna Freud did not expect Bowlby to change his position, but she thought he would listen to and respect his critics. She encouraged Schur to treat Bowlby this way and have an open discussion with him about their differences.[71]

Bowlby and Schur eventually met in Chicago in early May 1961 and talked about their views. Afterward, each reported to Anna Freud that there was a total lack of understanding on the part of the other. Schur's letter recounted at length what they said to each other, mainly a reprise of what they had put into print, and noted that they could not find common ground. Anna Freud told Schur she had received a letter from Bowlby about the meeting. Bowlby reported that it was "cordial enough" but that Schur had "great difficulty in understanding the legitimacy of a point of view different from his own." Bowlby also complained that Eissler did not want to print his response. Anna Freud suggested to her that he had the right to a reply and wondered whether Schur would support her petition to Eissler. Bowlby's response was published.[72]

The controversy ceased in the public realm, but this episode had brought no rapprochement. As Schur wrote to Anna Freud, the "Bowlby affair" had led to nothing. However, it had made clear that major psychoanalysts found Bowlby's synthesis of ethology and psychoanalysis unacceptable.[73] That was also the view of the other child psychoanalyst besides Bowlby who was most interested in Lorenz's work: René Spitz.

René Spitz

Bowlby and Spitz shared a lot. As we have seen in chapter 1, they both focused on observational studies of children suffering from maternal separation and deprivation. They both concluded that the lack of mother love had profound emotional and psychological consequences for infants. They were both interested in biology. For example, both turned to embryology for guidance on development, and both adopted the concept of an organizer for the human mind. In *A Genetic Field Theory of Ego Formation*, Spitz recounted how Spemann had introduced this concept in embryology as a developmental pacemaker and said that the development of the embryo was guided by differentiation of tissue dependent on a sequential organizer. By analogy, Spitz presented ego development as guided by different organizers, with the first organizer being the smiling response.[74]

Furthermore, Lorenz profoundly influenced both of them. Spitz had become enthralled with ethology and its implications for understanding human development and social behavior. He said he had "known and admired" Lorenz since 1935, and they became friends about 1952. He probably met Lorenz through Katherine Wolf, a child psychologist who worked at the University of Vienna from 1930 to 1938. She was a friend of Lorenz and also coauthored some papers with Spitz. Many years later, introducing Lorenz for a talk in the United States, Spitz presented himself as following "respectfully in the trail he is blazing." He said animal psychology was a different discipline than psychoanalysis, but one that offered "a view from another window only."[75]

In the mid-1950s Spitz wrote about ethology and its significance for child studies. He saw himself doing ethological work in his observations of children. As I noted earlier, in 1959 he even spent a few months in Lorenz's lab in Seewiesen, Germany. In a speech he noted that some psychoanalysts, including himself and Heinz Hartmann, had "always maintained the importance of innate behavior patterns, automatisms, primary autonomous sectors of the ego."[76]

Like Schur and Anna Freud, Spitz disagreed with Bowlby on a number of smaller points of interpretation, but their main difference came down to the same issue: Bowlby's reduction of psychology to biology. Spitz criticized Bowlby for not differentiating among children of different ages, and for assuming that losing the love object (in this case the mother) causes pain, grief, and mourning in the same way as in an adult. But the point that Spitz raised over and over again was the need to distinguish between biological data and psychoanalytic concepts.

Spitz did not consider the biological data wrong, or even irrelevant, but he saw them as insufficient to account for the formation of emotional relations. Here is what he argued:

> One might say that these innate response patterns are necessary, but not sufficient, conditions for the formation of object relations. These earliest interchanges provide the child with experiences that partake at the same time of the physiological and of the psychological. They trigger the first psychological processes and thus endow object relations with psychological content and meaning. This process transforms the object relations: at their inception they were essentially biological and mechanical. But in the process of development they gradually and increasingly assume the nature of an interaction that is primarily of a psychological nature.[77]

If biological processes and behavioral data were insufficient to account for the formation of emotional attachments, they were also not enough to explain the catastrophic consequences that Spitz, like Bowlby, believed maternal deprivation entailed for the infant. Spitz noted that Bowlby explained the syndrome of anaclitic depression and hospitalism by the "rupture of a key relationship and the consequent intense pain of yearning." Spitz agreed with this description of the observable phenomena, but he found it insufficient as "an explanation in terms of underlying dynamics."[78]

As Spitz emphasized, Bowlby limited himself to the behavioral level and thus had eliminated the psychoanalytic structural viewpoint, the division of the psychic apparatus into id, ego, and superego, and the dynamic viewpoint, which explains the formation of relations by appealing to aggressive and libidinal energies. Spitz even said that there was nothing wrong with this, but it simply was not psychoanalysis: "That is a legitimate approach and, like other psychoanalytic authors, I have endeavored to follow it for the observational and descriptive part of my work. But for the purpose of explaining empirical data in terms of psychoanalytic theory, this approach is inadequate."[79]

Although Spitz noted his use of ethological "methods" and biological concepts, his interpretation of the effects of children's separation from their mothers was embedded in a psychoanalytic framework. For him, the emotional deprivation children suffered in institutional care led to a lack of integration of the ego. What Anna Freud had called the "metapsychology" was necessary on top of the biology.

Conclusion

Bowlby was probably right in stating that the objections of Anna Freud, Max Schur, and René Spitz acted as red lights that kept younger psychoanalysts from accepting his work. Their seniority and standing in the field must have carried some weight. But most important, their arguments made it very clear that there was no possible synthesis of ethology and psychoanalysis in the terms Bowlby proposed. Even child psychoanalysts with views very similar to Bowlby's, such as Spitz and Winnicott, rejected Bowlby's assimilation of ideas from animal behavior.[80] The differences were not a matter of interpretation of specific cases or disagreements about the weight of any given factor. The differences were fundamental. They pertained to what level of analysis is relevant in explaining mental and emotional development and what tools are needed to study these matters.

For Sigmund Freud, one could not reduce human needs and motives, not even the child's love for the mother or the mother's love for the child, to a matter of biology. Freud never clarified the role of biology in his framework, but he would never have reduced mental life to a mere epiphenomenon of biological processes. If biology alone could explain human behavior and emotions, there would be no need for psychoanalysis.

But the varied objections Bowlby's critics presented were not simply a defense of psychoanalytic orthodoxy. This was not a dispute between those who valued facts (Bowlby) and those who defended psychoanalytic concepts at all costs, or between those who understood biology (Bowlby) and those who did not (his critics).[81] The questions concerned which facts and which concepts can help explain the human mind and its pathological conditions. The dispute was about fundamental questions: Can the human mind be completely explained by biological phenomena? Can psychoanalytic instinctual drives be reduced to biological instincts? What are the tools that allow us access to the complex world of human motivations and frustrations? Bowlby and his opponents gave radically different answers to those questions.

Whereas Bowlby was willing to use only behavioral data to make inferences about mothers' and children's inner emotions and psychological needs, Anna Freud and Spitz pointed out that those observations were insufficient to understand internal mental processes. In addition, Anna Freud pointed out that observations of children in hospitals were not adequate to support Bowlby's views about the negative impact of maternal separation in normal circumstances and that what affected a child negatively was the manner of separation. Whereas Bowlby adopted Lorenz's concept of instinct, Schur pointed out that the notion of preprogrammed behavior patterns eliminated the important role of experiences in shaping human mental states and emotional needs. Bowlby should not have been surprised by their reactions: his proposal for a synthesis of psychoanalysis and ethology fully eliminated the psychoanalytic structural and dynamic viewpoint of the human psyche.

CHAPTER SIX

Primate Love

Harry Harlow's Work on Mothers and Peers

Introduction

"Love is a wondrous state, deep, tender and rewarding." But because of its intimate and personal nature, scientists had considered love to be "an improper topic for experimental research." Thus in his 1958 presidential address to the American Psychological Association did Harry Harlow express concern about his fellow psychologists' lack of interest in a motive that "pervades our entire lives." In his view, psychologists were failing in their mission "to analyze all facets of human and animal behavior into their component variables." Harlow's talk, "The Nature of Love," changed the status of mother love within the laboratory and beyond. His experiments involving rhesus monkeys with surrogate cloth and wire mothers have become legendary in the scientific community and in popular culture.[1]

John Bowlby and later Mary Ainsworth presented Harlow's work as providing crucial empirical support for their ethological theory of attachment behavior. As we have seen, this theory postulated that children have an instinctual need for maternal care and that disruption of the mother-infant bond has severe consequences for the infant's psychological development. They also saw Harlow's work as confirming their belief in a critical period of early development that determines an individual's adult character.

To this day, most presentations of Harlow's work have basically repro-
duced this account or taken it for granted. According to several commentators, Harlow's work on rhesus monkeys raised with substitute mothers
corroborated in nonhuman primates the views of psychoanalysts such as
John Bowlby and René Spitz, who identified maternal love as necessary
to a child's adequate emotional development. Many authors also present
Harlow as part of a triumvirate that included Konrad Lorenz and Bowlby.
In this group, Lorenz provided the theoretical foundation for explaining
the instinctive character of human behavior, Bowlby contributed observational work on children, and Harlow provided the experimental work that
corroborated the instinctual nature of social behavior in primates. As in
all animal fables, Harlow's monkeys thus revealed to humans a deep truth
about nature: thwarting biological instincts causes profound disturbances.
For those authors, the moral of this story is about the power of biological
drives and the determinant effects of mother love in early infancy. Other
historical commentators like Carl Degler and Matt Ridley have also presented Harlow's work as an important step in the rehabilitation of instincts in explanations of behavior.[2]

Here I examine Harlow's work, its reception among contemporary scientists and the public, and its relevance to the ethological theory of attachment behavior. This chapter is not a full exposition of Harlow's views
on emotions and instincts over his career. My analysis centers on whether
his experiments proved the instinctual nature of the mother-infant dyad.
I show that Harlow did not think his results supported Lorenz's ideas
about imprinting and instincts or Bowlby's views about the infant's innate need for mother love. Harlow's results, in fact, called into question
their emphasis on the power of innate biological needs to determine adult
emotions and behavior.

In addition, this chapter shows, contrary to some interpretations of
Harlow's work and its influence in American society, that Harlow's experiments opened a wide debate about using animals as models for human
behavior. Donna Haraway has argued that Harlow "could design and
build experimental apparatus and model the bodies and minds of monkeys to tell the major stories of his culture and his historical moment."
In her view, "the laboratory rhesus monkeys ... complied in the production of discourse in the rhetorics of their own pliable bodies."[3] The assumption underlying this interpretation is that the rhesus monkeys were
passive subjects, "discursive constructs" elaborated by the experimenter.
But as my analysis will indicate, they did not always comply with the ex-

perimenter's expectations or desires. Nor was transferring experiments into society a straightforward affair. Society was not a passive receiver of scientists' pronouncements. In sum, neither Harlow nor society nor the monkeys told a simple story. Above all, the monkeys' complex behavior would become a good reminder that primate social behavior cannot easily be reduced to a simple activation of predetermined instincts.

Harry Harlow

Harry Frederick Harlow was born in 1905 in Fairfield, Iowa, and raised in the Midwest and Southwest of the United States. Until 1930 he lived as Harry Israel, but though he was not Jewish, his advisers at Stanford University thought that in job applications his Jewish-sounding name would affect his chances for a good academic position. Harry Israel suggested a couple of possibilities to his adviser, Lewis Terman, who selected the family name Harlow.[4] After obtaining a PhD in psychology for his work on maternal behavior in rats, as well as a new name, Harry Harlow arrived in Madison, Wisconsin, during the early years of the Great Depression, to start his work on comparative psychology.

He found no animal laboratory and little sympathy for his effort to breed rats in a classroom below the dean's office. On a hint, Harlow paid a visit to the Madison Zoo and made what he claimed was his first important discovery: after working with primates, one could never go back to "rodentology."[5] Rats, he believed, could afford limited conclusions of human interest. Harlow used to comment that psychology had done much for rats and pigeons, but he wanted to do something for humans. Harlow first pursued his research on monkeys at the Madison Zoo and in several makeshift research spaces, until in 1954 an old cheese factory at the university was remodeled into the Primate Laboratory.[6]

By the time he started to study love in the mid-1950s, Harlow had already established himself as a leading comparative psychologist. He carried out most of his research with rhesus monkeys, *Macaca mulatta*. His early work focused on learning and motivation in monkeys, and he had also done extensive research on the cortical localization of intellectual functions and on radiation effects. From 1950 to 1952, he was the US Army's chief psychologist, and he helped create the Human Resources Research Office, a military social science research center in Washington, DC. In 1951 he became the editor of the *Journal of Comparative and*

Physiological Psychology. The same year, he was elected a member of the National Academy of Sciences. In 1958 he became president of the American Psychological Association. In his presidential address to the association he presented his recent work on the origins of love.

In Search of the Origins of Love: Contact or Food?

After problems with importing rhesus monkeys from India led scientists in the United States to start breeding their own, in 1956 the University of Wisconsin became one of the first research sites to establish a breeding colony.[7] Rhesus monkeys are very susceptible to tuberculosis, and the illness spreads like an epidemic, making it too dangerous for infants to stay with their mothers after birth. On one occasion Harlow's team lost almost the entire colony. Wisconsin's breeding colony was successful because researchers followed strict procedures to keep the monkeys isolated from each other. After an infant was born, usually at night, the mother would eat the placenta and clean the infant. The next day, the research team removed the infant from the mother. Scientists put the baby monkeys in individual cages, and their experimental lives began. For the next three decades, the staff of Harlow's lab recorded all details about the monkeys' bodies and behavior: body changes, menses, coitus, grooming, looks, moods, fights, and embraces.[8]

One of the first things caretakers noticed seemed very odd: the infants developed a strong attachment to the folded gauze diapers covering their cage floors. It was not surprising that, alone in a cage, a baby monkey would grasp the only soft thing available. But without the diapers, the rhesus babies often did not survive.[9] This was strange, because wild rhesus monkeys live in a wide range of environments, from deserts to forests, mountains to swamps, villages to crowded cities. Why did they die in great numbers in a laboratory where they had all the comforts of modern life—controlled temperature, regulated diet, and constant care? Could a blanket make the difference between life and death? What caused the monkeys' attachment to the blankets, which they hugged, soiled, and held on to for dear life?

Thus Harlow was drawn to examine infant development in monkeys, and more specifically their social and emotional development. Harlow had written his dissertation on maternal care and had taught and written about emotions as early as 1932, but it was not until the mid-1950s that he turned to the experimental analysis of love.[10] At this time, as I have shown in earlier chapters, psychoanalysts and psychologists had given much attention to explaining the origins of love. In animal research, ethologists and comparative psychologists debated whether instincts determine social behavior.

As the editor of the major journal in comparative psychology, Harlow was aware of the controversy between comparative psychologists and ethologists about the role of biological instincts in animal and human behavior. On January 5, 1957, he accepted comparative psychologist Frank Beach's invitation to participate in a seminar on ethology and psychology, to be held that summer at the Center for Advanced Study in the Behavioral Sciences at Stanford University. This key meeting of major ethologists and comparative psychologists, including Daniel Lehrman and Robert Hinde, sought to create a rapprochement between the two communities. Harlow later wrote to the director of the Center, Ralph W. Tyler, to say how much he enjoyed the conference and the opportunity to present his "Babyhood and Motherhood project." Harlow did not specify what his project was, but he was probably referring to his latest research on infant rhesus monkeys raised with dolls as surrogate mothers.[11]

Hinde told Bowlby about Harlow's work. Bowlby immediately wrote to Harlow on August 8, 1957, then mailed him a draft of his paper "The Nature of the Child's Tie to His Mother." Pointing out their convergent interests, Bowlby requested copies of Harlow's papers and said he would like to visit him the following year, when he would be a fellow at Stanford's Center for Advanced Study in the Behavioral Sciences. Harlow extended an invitation to Bowlby, and a few months later he requested more copies of his paper. Bowlby visited Harlow's lab on June 15–18, 1958. From this point on, Harlow and Bowlby maintained contact through visits and conferences.[12]

The following year, Harlow and his student R. R. Zimmerman put forward their first results on affectional systems in rhesus monkeys as a contribution to the nature-nurture debate concerning emotions and behavior. Harlow and Zimmerman also situated their work in this area within the context of current research on the mother-infant dyad, citing Lorenz's work on imprinting and the work of child psychoanalysts, including Bowlby, which they cited as "personal communication."[13]

As reported in his 1958 presidential address, Harlow aimed to test "the relative importance of the variables of contact comfort and nursing comfort" in the formation of an infant's love for its mother. For Harlow, con-

temporary theories postulating that infants' love for their mothers grows
out of their appreciation for the nourishment they provide were unsatisfactory for two main reasons. First, they failed to account for the persis-
tence of infant-maternal ties after the mother stops providing food. Second, those theories did not explain how love for the mother becomes the
wellspring of love for other individuals. As an exception to that model,
Harlow referred to Bowlby, who, according to Harlow, "attributes importance not only to food and thirst satisfaction, but also to 'primary object-
clinging,' a need for intimate physical contact, which is initially associated
with the mother."[14] Yet Harlow noted that "as far as I know there exists
no direct experimental analysis of the relative importance of the stimulus
variables determining the affectional or love responses in the neonatal
and infant primate." Harlow argued that the human neonate would not
be a good experimental subject because by the time the infant develops
motor responses that can be "precisely measured," the determining conditions have been lost "in a jumble and jungle of confounded variables." His
solution was to use the infant monkey.[15]

Harlow's experiment involved raising infant rhesus monkeys with dolls
as surrogate mothers. These surrogates differed in the "quality of the contact comfort" they could supply. One surrogate made of wood was covered with sponge rubber and sheathed in terrycloth. A light bulb behind
her radiated heat. Harlow thought this surrogate was an ideal mother,
"soft, warm, and tender, a mother with infinite patience, a mother available twenty-four hours a day, a mother that never scolded her infant and
never struck or bit her baby in anger," whereas real mothers often failed
their offspring by breaking down emotionally and physically. In creating
a substitute, Harlow claimed he had "engineered a very superior monkey
mother," a "mother-machine with maximal maintenance efficiency, since
failure of any system or function could be resolved by the simple substitution of black boxes and new component parts." The other surrogate
mother was a similar "machine," but its body was made of wire (see fig.
6.2).[16] Harlow placed the cloth mother and the wire mother in different
cubicles attached to the infant's living cage. For four newborn monkeys,
the cloth mother lactated but the wire mother did not. For another four,
the condition was reversed.

The result? The monkeys took their milk from the mother with a
bottle, but they spent almost the whole day clutching the cloth mother
machine. Harlow concluded that the wire mother was "biologically adequate but psychologically inept." In his humorous way, he summed up his

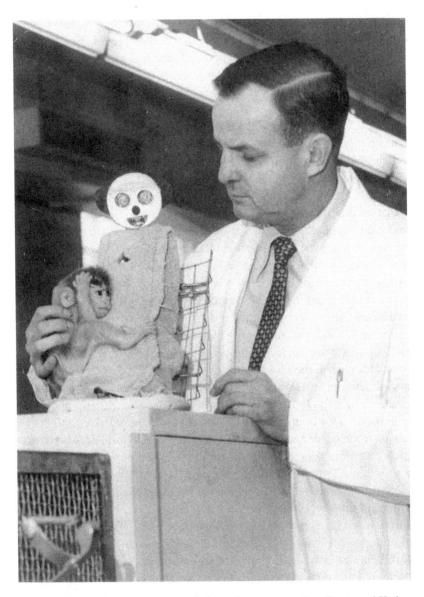

FIGURE 6.1. Harlow with infant rhesus monkey and surrogate mother. Courtesy of Harlow Primate Laboratory, University of Wisconsin, Madison.

FIGURE 6.2. Baby rhesus monkey with surrogate mothers. Courtesy of Harlow Primate Laboratory, University of Wisconsin, Madison.

interpretation: "The disparity is so great as to suggest that the primary
function of nursing as an affectional variable is that of insuring frequent
and intimate body contact of the infant with the mother. Certainly, man
cannot live by milk alone. Love is an emotion that does not need to be
bottle- or spoon-fed, and we may be sure that there is nothing to be
gained by giving lip service to love."[17]

Harlow drew larger conclusions about the mother's role in her child's
emotional development. He argued that mothers, "human or subhuman,"
help the infant feel safe and secure. According to him, it was evident that
the mother provides a haven because in times of fear, the "frightened or

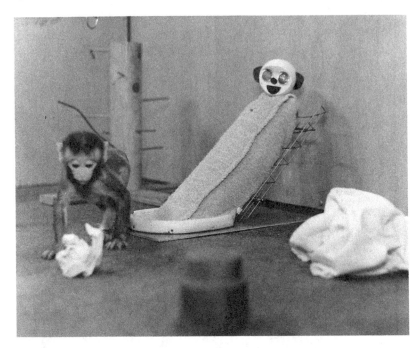

FIGURES 6.3. Object exploration in open-field test in the presence of the cloth mother. Courtesy of Harlow Primate Laboratory, University of Wisconsin, Madison.

ailing child clings to its mother, not its father; and this selective responsiveness in times of distress, disturbance, or danger may be used as a measure of the strength of affectional bonds."[18] Harlow, however, did not refer to observations or experiments that had assessed the different responses of infant monkeys to mothers and to fathers.

As for the view that the mother is also a source of security, Harlow presented experimental evidence drawn from his "open-field test" in the "strange room." Harlow put four of the original eight monkeys raised with dual surrogates into a room that contained objects known to elicit curiosity in baby monkeys. The infants were taken to the strange room twice a week for eight weeks, one time alone and the other time with the cloth surrogate. After a couple of adaptation sessions, the baby monkeys rushed to the cloth mother, clutched her, and rubbed their bodies against her. They used this mother surrogate as a base of operations, venturing away to explore and manipulate something in the room, then returning to her comfort before venturing forth again (see fig. 6.3). How-

The Elephant   Ylla

Though mother may be short on arms,
Her skin is full of warmth and charms.
And mother's touch on baby's skin
Endears the heart that beats within.

FIGURE 6.4. Elephant mother-infant pair, with accompanying poem. From Harry Harlow, "The Nature of Love," *American Psychologist* 13 (1958): 673–85, illustration on 678. Courtesy of the American Psychological Association. Reprinted with permission.

FIGURE 6.5. Harlow's experiments featured in a comic strip. In Harlow's archival papers.

ever, when the cloth mother was absent, the infants froze in a crouched
position, rocking, sucking, and frantically clutching their own bodies. The
same behavior occurred in the presence of the wire mother. There was no
difference between the infants fed by the cloth mother and by the wire
mother.[19]

Harlow also compared surrogate mothering with real mothering. He
reported observing the behavior of two infants raised by their own mothers. In his opinion, love for the real mother and love for the surrogate
mother were similar. In both cases there was "togetherness."[20]

To support his conclusions, Harlow presented powerful visual evidence: a twenty-minute movie revealing the love between infant and
mother. Words and images: Harlow knew the power of both, and he combined them in poems coupled with pictures of mother-infant pairs of rhi-
noceroses, snakes, elephants, and crocodiles (see fig. 6.4).[21]

"It was absolutely superb: the substance, the wit and the delivery. More
power to you," the eminent psychologist Edward Tolman wrote to him
after his 1958 presidential address to the American Psychological Association.[22] With the movie, the poetry, the jokes, and the explicit extrapolation from infant monkeys to human infants, Harlow's was no standard academic address. It became an instant success inside the scientific
community and outside it. His studies became widely known, and he received acclaim from many colleagues. His work was reported in *Newsweek*, the *New York Times*, *U.S. News and World Report*, and many regional and local newspapers. Even some of his poems were published in
the *New York Times Sunday Magazine*.[23] Over the years, the mother ma-

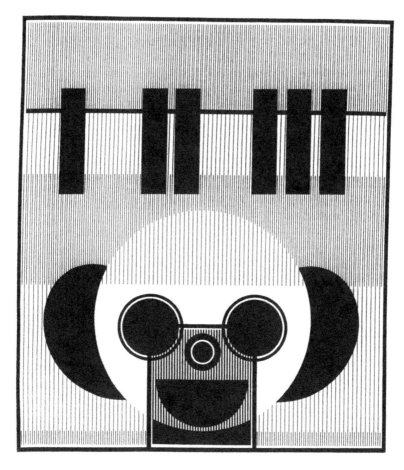

"The Three Programs of Dr. Harlow"—See page 4   *July–August 1966*   Number 27

FIGURE 6.6. Harlow's mother machine on the cover of *SK&F Psychiatric Reporter*, no. 27, July–August 1966. In Harlow's archival papers.

chine made it into science journals, art magazines, and comic strips (see figs. 6.5 and 6.6).

Sometimes modified, often misrepresented, always simplified, Harlow's experiments with surrogate mothers fired the public imagination and reached iconic status in psychology and popular culture. But what did Harlow's results about the infant rhesus monkey's need for contact comfort imply for contemporary theories of child development and views about child care?

The Machine (or the Father) in the Nursery

When he described the relationship between infant and mother monkeys as a clear expression of "togetherness," Harlow was appropriating a metaphor that sent a powerful message through American society during the 1950s. As historians have documented, amid widespread anxiety and insecurity in the increasingly tense climate of the Cold War, togetherness became the rallying call for a postwar social order based on traditional gender roles. After World War II, specific economic and social measures shored up the patriarchal family, reaffirming the man as provider and the woman as homemaker and mother. In addition, the rhetoric of family togetherness encouraged traditional parental roles.[24]

According to a number of social scientists, contemporary research also showed that the patriarchal family was the best arrangement for helping the mother provide the contact comfort her children need. As I explored in chapter 3, child psychoanalysts like Margaret Ribble, Spitz, and Bowlby urged mothers to stay home and devote themselves to raising their children. And along came Harry Harlow with experimental proof that babies need their mother's embrace.

Yet, making a significant substitution in the conclusion to his APA address, Harlow ended up talking not about mothers but about fathers:

> The socioeconomic demands of the present and the threatened socioeconomic demands of the future have led the American woman to displace, or threaten to displace, the American man in science and industry. If this process continues, the problem of proper child-rearing practices faces us with startling clarity. It is cheering in view of this trend to realize that the American male is physically endowed with all the really essential equipment to compete with the American female on equal terms in one essential activity: the rearing of infants.[25]

Thus Harlow argued that an artificial surrogate had proved that the father could be as good as the natural mother, in monkeys as well as humans.

Harlow's claims provoked an animated public debate. The long *New York Times* article covering his experiments brought a disturbing thought to the fore: "A 'mother machine' raises the question of how necessary is mother love." Could a machine take over a mother's place? Technological advancement was central to the new American postwar order, as Vice President Richard Nixon argued in his infamous 1959 "kitchen debate" with Soviet leader Nikita Khrushchev. Nixon claimed that the superiority of American democracy over other ways of life was marked by its technological progress. So if Harlow was right that a machine could provide the haven an infant needed, what would prevent the machine from moving from the garden into the nursery? Perhaps, like the infant rhesus monkey, all a child needs is to clutch a soft and warm doll with big eyes.

But Marston Bates, the zoology professor at the University of Michigan who wrote the *New York Times* article, explained that Harlow's apparatus was not really a mother machine, but a machine for postnatal care. He noted that postnatal care was a problem in the economy of nature, and nature had provided a solution for each species. Nature had also designed the best solution for humans. Unless society got a "psychologist to fix a machine," Bates asserted confidently, human mothers were stuck with their offspring.[26]

An adequate machine for human infant care was still in the realm of technological fiction, but Harlow had presented an alternative closer at hand: Dad. If, as Harlow claimed, contact comfort was all an infant needs, the message was clear. As a newspaper report put it: "*Anyone* could be a mother."[27] Furthermore, Harlow had made the equivalence between the machine and the father explicit. Thus the introduction of the machine into the nursery not only constituted a technological threat to the natural benefits of mother love; it also threatened to disrupt parental roles widely taken to be deeply rooted in nature.

In the press, the implications of Harlow's experiments became more and more threatening to the mother. According to a report in a Virginia newspaper, he had "concluded from experiments that fathers could rear infants, replacing mother love." In a Detroit newspaper, the news grew bleaker: "A University of Wisconsin psychologist has confirmed what most men suspected all along: that fathers make better mothers than the mothers do." The message: "Who is the best mom? It's dad, says expert." Thus, not only could fathers mother—they could do it better than real moms. According to this report of an address he gave at Florida State University, Harlow claimed, "The American male is physically endowed with all the really essential equipment to compete with the American female on equal terms in the rearing of infants." Therefore it was entirely possible that men would replace women at home.[28]

Some women greeted this possibility as a blessing. An editor who presented some of Harlow's results to her local cooperative nursery school wrote to him that the mothers were fascinated. They were "almost inde-

FIGURE 6.7. "And After Motherhood—What?" Courtesy of Harlow Primate Laboratory, University of Wisconsin, Madison.

cently delighted with your conclusion about fathers and child rearing." In her letter, this editor also included a special cover page designed by another woman who was impressed with Harlow's work. It depicted an attractive woman dressed in electric blue leggings and a flashy pink tank top, reclining with a long cigarette holder. A small monkey is perched on her neck. The caption read: "And After Motherhood—What?" (see fig. 6.7). Clearly, these women saw the possibility of liberation, thanks to Harlow and the rhesus monkey.[29]

Other women, though, found Harlow's conclusions almost offensive. In the context of the Cold War togetherness craze, with mothers busy in the suburbs, often bored and overworked but sustained by their confidence that their job at home was essential for the well-being of their children and society, some did not welcome the view that "anybody" could mother. One of these busy women took her concerns to the top of the academic ladder. "Raising monkeys and raising children have no comparison," a mother of eight children wrote to the president of the University of Wisconsin.[30]

This mother had put her finger on a key issue: Could one extrapolate from the infant rhesus monkey's need for contact to the human baby's need for love? In the *New York Times*, Bates had also argued that experiments with monkeys were not relevant for deciding how to raise human babies. In the *Springfield (OH) News-Sun*, journalist Maynard Kniskern accused Harlow of moving recklessly across too many boundaries. "'Infant macaque monkeys . . . have the same feelings of affection as humans.' This is a wildly unscientific statement," he claimed. He continued, charging that "Dr. Harlow hasn't the faintest idea of whether infant macaque monkeys experience 'feelings of affection,' much less whether what they feel is in any manner akin to what human infants feel." In Kniskern's view, Harlow's experiment could reveal little or "nothing about human babies. It doesn't even tell us much about baby monkeys. We still don't know how infant monkeys 'feel' about mothers: we simply know how a few of them react to an assortment of dolls fabricated by Dr. Harlow and his aides."[31]

These criticisms focused on two crucial questions about Harlow's work: In observing the rhesus monkeys' need for contact comfort, was Harlow witnessing their need for love? And do humans have the same affectional needs as rhesus monkeys?

Regarding the first question, Harlow was clearly after love and thought he was getting at it, though others found his optimism unjustified. Harlow believed that rhesus monkeys would lead him to a complete explanation of love. He confidently claimed, "Indeed, the strength and stability of the monkey's affectional responses to a mother surrogate are such that it should be practical to determine the neurological and biochemical variables that underlie love."[32] But some critics argued that he was simply measuring the infant monkey's proximity to the cloth mother. Harlow retorted by saying that perhaps proximity was all his critics knew of love, but he was grateful that he had known more.

In Harlow's experiments, what was this thing called love? Sometimes Harlow defined love as a complete sense of security, while at other times he said love was crucial for the development of security. His experiments showed that the infant monkeys in the strange room seemed relaxed when the cloth mother was present, whereas in her absence they were terrified. But the precise relation between love and security remained unclear. A sense of security was greatly desired in American society during the Cold War. If keeping close to one's mother could provide that, perhaps it did not matter whether that was the cause or the result of love.

Harlow knew that nobody is interested in how close a baby monkey needs to be to its mother. Many more people are interested in security. And in love, who isn't?

Still other critics insisted that Harlow might not even be observing the monkeys' need for security, but only the creation of fetishism. Psychoanalyst Erik Erikson, a great admirer of Harlow, confessed he believed that Harlow's experiments had little to do with love:

> Dr. Harlow, in a way which is admittedly humorous and analogistic, uses such phrases as mother-love in connection with experiments which in themselves, of course, are most ingenious. Maybe I am a sentimental clinician who cannot accept the idea of calling a wire cage a mother. But I cannot help thinking that there is probably something scientifically wrong in this nomenclature. To my mind, all the functions of a mother animal in the life of a monkey are here taken care of by human beings.... [T]o me, these are experiments in creating fetishism in monkeys.[33]

Perhaps the mother machine was not an object of love, but a fetish that attracted monkeys and, others argued, even humans. Responding to a laudatory review of Harlow's presidential address in the *American Psychologist*, Solomon D. Kaplan of the Lincoln State Hospital in Nebraska asserted that Harlow's surrogate doll served as a fetish for humans. To study the nature of patriotism, Kaplan jokingly suggested imitating the Stars and Stripes to create a surrogate monkey flag with crossed bananas on a background of peanuts![34] Proximity, security, fetishism, ... what exactly did Harlow's experiments reveal about the needs of infant monkeys?

The second crucial question raised about Harlow's work was whether his results were relevant to humans. In his earlier research, Harlow showed that monkeys are curious, as we are, and that they learn to think, as we do. But do they love as we love? Harlow often joked about extrapolating from monkeys to humans. In his zesty way, he quipped that monkeys are not little people with tails, and "men are not monkeys most of the time."[35] Yet he had no doubt that one could extrapolate:

> It is my conviction that one cannot directly generalize from monkeys to man, but I believe there are many parallels in the psychosocial development of monkeys and the psychosocial development of human beings. I believe that our data on rearing of infant monkeys under controlled and varied social conditions illuminate much human data on both normal and disturbed children.[36]

Harlow even provided a numerical conversion table: "To find a rough developmental equivalent for human babies, one multiplies the monkey's age by 4 or 5." Furthermore, in many presentations of his findings, he claimed the existence of the same basic affectional systems in all Anthropoidea: monkeys, apes, and humans. He joked that animal psychologists thought of themselves as theoretical psychologists because their obligation was "to discover general laws of behavior applicable to mice, monkeys, and men."[37]

It is also important to note that the agencies supporting Harlow's research were interested in the human implications of his studies, as became clear when public controversy erupted over the wisdom of providing federal support for his work. The debate ranged from west to east, from the humor pages of newspapers to prestigious scientific journals like *Science*, from the US House of Representatives to the Senate. The first shot came in 1962 from Senator Harry Byrd. The press turned it into a national affair, as the *Arizona Election News* alerted the public that $50 million of their hard-earned tax money had been used to "pay for studies to find out if baby monkeys love their mothers."[38] The alleged amounts varied in different newspapers, as did the assessments of Harlow's research. In one cartoon the issue was put forward with graphic clarity. A monkey is represented asking another monkey nursing an infant about the national debt. The caption read: "The new Frontier plans to spend $1,201,925.22 to study the affection of the monkey's offspring."[39]

When Harlow's lab started experiments on affectional systems in 1956, a healthy infant monkey was worth about $1,500. In the following ten years, the lab experimented with hundreds of monkeys. To cover growing expenses, it received support from a variety of sources: the Wisconsin Alumni Research Foundation (WARF), the Graduate School of the University of Wisconsin, the National Institutes of Health, the National Heart Institute, the Division of General Medical Science, the Ford Foundation, and the Department of Defense. Toward the end of his stay at Madison, Harlow stated that he had received over $10 million in the previous ten years alone. Although not all that support went to the study of affectional systems, clearly the search for the nature of love was not cheap. The support Harlow received provides one indication of how many people found it worthwhile.[40]

The interesting point for us here is that all who wrote in Harlow's defense did so because of his experiments' perceived relevance to humans. In response to an editorial in the *Chicago Daily News* that criticized Harlow's work as trivial, a group of scientists from the University of Illinois rose to vindicate it. Outraged as much by the arrogance of outsiders who assumed they could assess the significance of scientific research as by the specific criticisms of Harlow, they pointed out that "his findings are so relevant to the comprehension of human psychopathology that one of us has incorporated them into the neurophysiology lectures to our medical students." A psychiatrist wrote that he regarded Harlow's work as "an exceedingly valuable area of investigation for giving clues and leads which may be of direct benefit to human beings."[41]

When elected officials had to debate whether the country should spend public funds to find out if infant monkeys love their mothers, Harlow's defenders followed the same strategy. In response to Massachusetts Democratic representative Edward P. Boland's inquiry about a grant from the National Institutes of Health given to support research on affection in monkeys, Wisconsin Republican representative Melvin H. Laird rose to defend Harlow. He noted that Harlow's work had been lampooned as a study of monkey business, but the issues were serious: "This caricature is all in good fun until it is taken seriously and used as a weapon to attack appropriations for medical research." His defense of Harlow rested on a detailed presentation of his scientific credentials and on drawing out, one by one, the many implications of his research for human problems. In a statement prepared to respond to public criticisms, the National Institute of Mental Health also noted that it supported Harlow not because it was "especially interested in the study of monkey behavior itself," but because it was interested "in the implications which these studies may have for a deeper understanding of human behavior."[42]

In sum, there is no question what Harlow and those who supported his work—from colleagues to funding agencies to many among the general public—were after. For them, the monkey was an instrument for understanding the human; and proximity was just a means to love.

But contemporary commentators disagreed about what Harlow's team had discovered regarding love or proximity in rhesus infants, and they argued about the implications of those results for humans. The experimental work with monkeys did not tell a uniform and straightforward story. The only noncontroversial result at this time was that, to survive in the laboratory, infant monkeys needed contact with something soft and warm. For Harlow and many others, it seemed quite reasonable to assume that human infants would also need some kind of contact-comfort. And, as Harlow had argued, perhaps fathers could provide that. However, scientists did not speak with a single voice about this issue.

Just as significant, in trying to assess the implications of Harlow's experiments for human infants and for child care, society was not a passive receiver of scientific knowledge. The implications that his results had for child rearing were a matter of considerable debate. The translation of the significance of his results with infant monkeys to human infants was open to a variety of interpretations. Therein lay their (controversial) appeal.

Baby monkeys and baby humans have at least one thing in common: they grow up. While scientists, politicians, and mothers were debating the needs of rhesus monkeys, the infant monkeys became young adults. But they did not behave as such.

The Machine Produces Monsters:
Bring Back Natural Mother Love

When the monkeys "mothered" by the machine surrogates grew up, they seemed very troubled. Males and females were paired on reaching maturity, but they showed little interest in each other. It turned out that the twenty-four-hour welcoming-mother machine was not so good at raising infants to become "normal" adults. In a sense she was too good. She never pushed the infant away from her warm contact. But what originally appeared to be a virtue turned out to be a vice. Too much mother love— or too much proximity to the mother—was as bad as too little, or even worse.[43]

To explain these results, Harlow identified two stages in good mothering. In the first, the mother provides contact, comfort, and security. The mother machine had passed this test with flying colors. In the second stage, the mother gradually relaxes the bonds between herself and the infant by pushing the baby out into the world, toward peers and play. According to Harlow, the patterns of childhood play are gradually overridden by the aggressive and sexual patterns. Play is an essential precursor to other social interactions. Without it, these subsequent patterns of adolescent and adult behavior will not develop normally.[44]

Harlow tested this hypothesis by creating a "social-mother situation" in which each of a pair of monkeys was raised with its own cloth mother surrogate while it was also free to make contact with another baby monkey. The data reported were "favorable to our theory that prolonged fixation to the mother surrogate adversely affects the subsequent capability of forming adequate infant-infant social contacts." Deprived of play, monkeys remained infantile. They all became physiologically sound, healthy

FIGURE 6.8. Rhesus monkeys in incorrect mating posture. Courtesy of Harlow Primate Laboratory, University of Wisconsin, Madison.

adults, but they would just sit staring into space, apparently uninterested in other monkeys or mates or incapable of relating to them. When females were paired with experienced males, the males tried to copulate but could not, since the females simply sat down. The males who were raised by mechanical mothers were equally ineffective (see fig. 6.8).[45]

Could there be any clearer sign of maladjustment than lack of interest in heterosexual sex or inability to perform what is commonly viewed as a natural, instinctive activity for which individuals need no training? Actually, for a female the answer is yes: lack of interest in her own infants. Harlow impregnated some females raised by the mother machines through artificial insemination. When they gave birth, these "motherless mothers" had no interest in their babies. In fact, they ignored or abused them. If the caretakers had not taken the babies away, most mothers would have killed them (see fig. 6.9). Thus the mother machine, source of constant and unconditional contact comfort, produced monsters: sexually incompetent males and infanticidal females.

The press quickly reported the awful new results: "Monkey Mothers Upheld in Tests: Imitation Parents Found to Foster 'Hopeless and Help-

FIGURE 6.9. Abusive mother. Courtesy of Harlow Primate Laboratory, University of Wisconsin, Madison.

less' Neurotics." As in a medical trial, the value of the mother machine could be assessed only in the long run. According to the *New York Times*, the results were now in, "in favor of real mothers." Several newspaper articles explained how "strange things" had happened since the first presentation of the mother machine. "The Troubled Monkeys of Madison" made it evident that mother love could not be dispensed with so easily. The machine was no substitute for the real mother.[46] Nobody said what that implied for fathers. But it seemed that mother—the natural one of flesh and blood—was here to stay.

In the scientific community, René Spitz saw Harlow's results as confirmation of his belief that maternal care and love are necessary for an infant's well-being. Spitz had heard Harlow present his work at a symposium of the American Association for the Advancement of Science titled "Expression of Emotions in Animals and Men" and spoke with him briefly afterward. In 1961, after seeing a report of Harlow's latest results in the *New York Times*, Spitz wrote to him, saying, "I am not sure you are aware of my work on emotionally deprived infants. If you are, you probably realize that your work comes as extraordinarily welcome confirmation of everything which I have found in the human infant, when placed in similar circumstances." Harlow answered that Spitz's work had actually encouraged him to study the effect that inadequate early affectional patterns had on subsequent behaviors. And in carrying out this research, he was struck by the "basic similarity of our monkey 'syndromes' and the syndromes described by you for affectionally deprived human infants." From then on, Spitz and Harlow kept in contact.[47]

John Bowlby used Harlow's results to support his view that separation from the mother has disastrous consequences for infant development. In a 1962 WHO report reviewing Bowlby's views about the devastating effects of maternal deprivation, Mary Ainsworth also noted that Harlow's experiments supported Bowlby's ideas, especially Harlow's finding "that monkeys which had been separated from their mothers at birth and reared either in the absence of any mother-surrogate or with only an inanimate cloth mother-surrogate failed, at maturity, to show normal sexual behaviour" and displayed "an abnormal absence of maternal behaviour."[48]

Harlow was aware that psychologists and psychoanalysts blamed a variety of conditions on the disruption of the mother-infant dyad. He knew those claims well: "Personality malfunctions that have been attributed to maternal inadequacy include such syndromes as marasmus, hospitalism, infantile autism, feeble-mindedness, inadequate maternal responsiveness, and deviant or depressed heterosexuality." Furthermore, Harlow often claimed that his work on monkeys could contribute to understanding those conditions.[49]

For example, in one of their experimental series, Harlow and his students Bill Sears and Ernst Hansen studied separation of infant rhesus monkeys from their real mothers. They noted Bowlby's theory of primary separation anxiety, which attributes separation anxiety to activation of the component instinctual response systems underlying the infant's attachment to the mother, in a situation where the mother is not present. Bowlby postulated that the infant goes through three phases: protest, despair, and detachment. Harlow's team argued that if those were basic biological mechanisms that activate instinctual responses, then a similar syndrome would appear in monkeys. The team tested this view by separating two pairs of infant rhesus monkeys from their mothers. Comparing their results with Bowlby's views about separation of human infants, they concluded: "There is great similarity in the variables which Bowlby (1958) has described in his Component Instinctual Response theory and the variables producing the infant monkey's tie to its mother."[50]

Harlow declared that the findings of his group's continuing experimental work on mother-infant separation were "generally in accord with Bowlby's theory of primary separation anxiety as an explanatory principle for the basic primate separation mechanisms." After separation, the mothers seemed emotionally disturbed. The infants were devastated, showing signs similar to the first two phases Bowlby described in humans: protest and despair. The third phase Bowlby described—detachment— and the signs of aggression against the mother did not appear in infant rhesus monkeys. Overall, however, Harlow's team gave the imprimatur to Bowlby's findings: "The results of this investigation appear to be in general accord with expectations based upon the human separation syndrome described by Bowlby."[51]

In the discussions of Harlow's results in conferences and the press, one response suggested they also fit within the imprinting framework that Lorenz had popularized in the United States. For example, after a report of Harlow's latest results in *Newsweek*, one reader, James H. Middlekauff, wrote to Harlow "surprised that nowhere in the article was the fact mentioned that perhaps the monkeys' odd behavior could be attributed to imprinting." Harlow replied, "I have just returned from a conference that was set up by the Menninger Clinic with one primary purpose, to bring Dr. Lorenz, who is a visiting professor there this semester, and me together. Dr. Lorenz had previously visited my Laboratory but I was out of town at the time."[52] Harlow was referring to a workshop titled "Approaches to Instinctive Behavior" that took place during January 1961 at the Topeka Institute for Psychoanalysis in Kansas. Anthropologist Margaret Mead and psychoanalysts Mortimer Ostow, Frederick Hacker, Karl Menninger, and Gardner Murphy also participated.

In his letter to Middlekauff, Harlow noted that it was "entirely possible that it is merely a terminological matter whether one wants to describe the kind of phenomena that we have demonstrated for our monkeys as imprinting or not." Yet Harlow asserted that imprinting was not an adequate mechanism to account for emotional development in primates: "However, formation of appropriate behavior patterns relating to the various affectional systems is not bound by such sharply circumscribed temporal periods as is the case of the birds and the fishes—the phenomenon to which Lorenz gave the name imprinting." In Harlow's view, imprinting was "probably not the mechanism in the monkey. Imprinting is not operating, obviously, in the infant-mother separation patterns, either for the monkey or for the bird."[53] Thus, for Harlow, even if imprinting was responsible for an infant's attachment to the mother in some species, it was not sufficient to explain an infant's ability to forge relationships with others. If an infant remained attached to the mother, as his infant monkeys remained attached to their surrogate mothers, it will be unable to engage in social relations with other infants and adults later on.

Harlow began to place greater emphasis on the need to explain not only how an infant attaches to its mother, but also how it then moves away from her to interact with other individuals and develop emotions for them. After all, a baby cannot retain an infantile attachment to its mother for the rest of its life: "An infant monkey cannot form adequate affectional patterns for other monkey infants unless it can break the contact bond which has been established between it and the mother."[54] For Harlow, Bowlby's and Lorenz's accounts did not explain how love for the mother eventually leads to love for other individuals. Exactly how does an infant generalize, transform, or divert its love for its mother to other members of the species? In addition, Harlow's monkeys soon proved more resilient than expected. As he proceeded with his experiments, Harlow separated himself from "the general accord" about the infant's need for mother love.

The Power of Peers

Harlow's infant monkeys had been deprived of more than their mothers' arms. The adult males and females who did not perform sexually and the mothers who did not provide maternal care had enjoyed no mother love in their infancy. But they had also been isolated from peers and grown up in an impoverished social environment. Harlow knew that their bizarre behavior could result from several types of deprivation: "We cannot be sure, of course, whether their failure to show normal maternal behavior stems from their motherless (or inadequately mothered) infancy, from their lack of association during the first years of life with other infants and young monkeys, or from both factors."[55]

What do infants really need: a mother, peers, neither, or both? What is really essential for their normal development? In a new series of experiments, Harlow's lab raised infant monkeys in conditions of total and partial social deprivation. Harlow aimed "to determine the effects of various kinds of social rearing conditions on subsequent social, sexual, and maternal adjustment of monkeys." In a 1962 paper he reported that monkeys raised in isolation from birth to eighty days, six months, one year, and two years were unable to form effective social relationships. He noted that "the data might also suggest that real monkey mothering is a variable of overwhelming importance for the normal psychosocial development of rhesus monkeys," but he cautioned against a hasty conclusion, because those monkeys had been deprived of both real mothers and infant-infant interactions.[56]

To ascertain the impact of separation from other individuals, Harlow carried out experiments to compare infant monkeys raised with a mother but without friends and infants raised without a mother but with friends. The lab researchers reared eight infants on cloth surrogate mothers and allowed them to interact in groups of four in specially designed play-rooms. They found that "for the most part" these monkeys developed normal play patterns, grooming, and sexual responses. At this point, Harlow recognized that they did not know whether real mothers made a "greater positive contribution than a surrogate monkey mother to monkey mental health if the infants are given even restricted opportunity to form affectional relationships with each other throughout the first year of life." But from his experience raising the infants of eight "motherless mothers" who had become inadequate (four), indifferent (two), or overtly abusive (two) mothers, he reported that when their infants became a playpen group, their play responses were close to normal and the infants were sexually precocious but otherwise normal. He concluded that, with the help of their peers, the "almost totally evil mothering" those infants received had "left no really serious deficits."[57]

Harlow reported two further experiments. In one, three females and one male were housed together and allowed access to the playroom. At one year, they were fully mature heterosexually and showed no social deficits. In another experiment, they raised one male and one female without access to other monkeys. At seven months they separated them from their mothers and exposed them to each other in the playroom. By one year of age, however, those monkeys had not developed "play behavior beyond that characteristic of young infants nor have they displayed any sexual behavior."[58] The experiments led Harlow to a startling conclusion: "In the monkey, at least, it would thus appear that under favorable circumstances, real mothers can be bypassed but early peer experiences cannot. Thus, when playmates were denied, the infant monkeys were socially crippled, and when this variable was provided early, the infants survived both passive and brutal mothering and even no mothering at all."[59]

FIGURE 6.10. Together-together rhesus monkeys. Courtesy of Harlow Primate Laboratory, University of Wisconsin, Madison.

Harlow did not deny that it was better for monkeys to be raised with their mothers. But he believed mothers were not absolutely necessary, and they were certainly not sufficient.

Harlow's next step was to see why those monkey mothers that had been raised with the surrogate mother machines and thus had no "real monkey mother of their own and also no opportunity to interact during childhood with other monkeys," became "hopeless, helpless, heartless mothers." What he now called the "pseudomothers" had appeared in principle to be very good mothers, since they never rejected their infants and provided them with food and contact comfort. But as they had discovered, their infants remained immature and socially inept. Now Harlow established an experimental setup for comparing how young monkeys would interact with real mothers and surrogates both in the absence and in the presence of peers. The results showed that real mothers facilitated the interaction of their infants with peers because at a certain point they encouraged and even forced their infants to move away from them.

Thus he concluded that an important function of the mother is to provide a bridge to the world of friends: "Real mothering is not a necessary condition for normal monkey social development although it may facilitate infant-infant interactions."[60]

Harlow, once again, did not deny the mother's significance but he rejected the view that the mother, and the mother alone, could guarantee adult adjustment:

> No one will question that real, normal monkey motherhood is an important variable in imparting normality to rhesus monkeys as revealed in social adjustment in infancy, adolescence, and adulthood. Furthermore, real monkey motherhood is doubtless a variable of far more importance in the wild than in the protected situation of an experimental laboratory. Nor would anyone seriously question that normal human mothering is an important variable in the social development of the child. Even so, these researches attest to the enormous importance of affectional relations between infants and preadolescent peers. In the monkey, at least, it would thus appear that under favorable circumstances, real mothers can be bypassed.[61]

Perhaps wary of changing his earlier position on such a delicate issue, he hedged his conclusion, adding (in another publication) that "at the present time, we tentatively conclude that adequate peer experience can compensate for lack of real mothering, and we hypothesize that real mothering cannot compensate for lack of peer association."[62]

In sum, Harlow's work indicated that it is best to have both mother and friends. But if one has to choose, peers are better able to compensate for the lack of mother than the other way around. Harlow concluded:

> The data from the isolation experiments and from the various experiments providing differing experiences with mothers or peers lead us to conclude that peer experience during early development is the sine qua non for adequate adolescent and adult monkey behavior. In spite of no mothering, surrogate mothering, or indifferent or brutal mothering, monkeys given regular opportunity to associate with peers from early in life develop behavior similar to or indistinguishable from that of monkeys provided the luxury of both normal mothering and normal peer experience.[63]

Harlow agreed with psychologists who said early social deprivation had profound consequences for later behavior. In further experiments subjecting infant monkeys to various degrees of social isolation, Harlow showed that deprivation produced bizarre social behavior. Monkeys raised in isolation or subjected to it for a long time were deprived of various cognitive stimuli and emotional interactions that affected their development in profound ways. Extreme isolation was so disturbing for infant monkeys that they were later unable to engage in any form of social relations.[64]

But on the central question that child psychologists and psychoanalysts raised about the role of mother in the emotional development of the infant, Harlow concluded from his experiments in the early and mid-1960s that the infant monkey could make do with peer love. As he put it, "The combination of mother and peers is the most advantageous for personality development. . . . But, and this is most important, monkeys deprived of real mothers nevertheless develop normal personalities if they are provided with peers for regular interaction."[65] This did not mean that mothers are irrelevant. But it did mean that those unlucky ones who grow up without a mother can survive well with other types of love.

Harlow was not shy in broadcasting his revised conclusions about the role of mothers and peers in the development of rhesus infants. As soon as his first experiments in this area were carried out, the results made it into the *New York Times*. According to an article titled "Child Adjustment Linked to Friends," Harlow declared that a mother's role might be secondary, or even dispensable, in her child's social adjustment.[66] In the next couple of years he boldly pointed out that his results contradicted common psychoanalytical explanations: "Our present conclusion is that in spite of completely inadequate, even brutal, mothering, the infants developed normal affectional behavior toward peers and normal heterosexual behaviors."[67]

The idea that maternal care and love are not the essential ingredients for an individual's normal emotional and social development contradicted standard explanations provided by human researchers such as Bowlby and Ainsworth. As Harlow noted: "This finding contrasts with current psychiatric and psychoanalytic theory stressing the importance of the mother's role and minimizing the part played by interactions among peers in the development of the normal adult personality."[68]

After his experiments with infant monkeys raised with peers, Harlow elaborated a conception of infant development that included a rich emotional universe and a complex conception of love. Over the years Harlow came to the conclusion that love is multifaceted. He and his wife, Margaret K. Harlow, proposed the existence of multiple affectional systems, first

speaking about the infant-mother affectional system and gradually adding four more: between infants or juveniles, heterosexual, mother-infant,
and father-infant. They argued that love does not have a single source but
develops from many systems, each one building, adding to, or compensating for the others. In their view, just as age-mate affection partially com-
pensates for inadequate mothering, so does maternal affection partially
compensate for inadequate peer socialization. Therefore it is inappropriate to pit one affectional system against another in importance. The existence of diverse systems that can compensate for each other provides an
enormous social safeguard, since both mothers and age-mates may be deficient or unavailable.[69]

The existence of several "affectional systems" made sense from an evolutionary perspective, the Harlows added. If love is necessary for survival,
it seems too risky to rely on one affectional system. In primates, socialization is essential to survival, and the hazards to successful socialization are
many. Harlow argued that the biological utility of compensatory social
mechanisms was obvious. That effective social safeguards should have developed over the course of evolution was in no way surprising. To prevent
the disastrous consequences implied by a failure of one system, evolution
had instituted compensatory systems. These are not surrogate mothers,
but surrogates for love: "From an evolutionary point of view there is gain
in having two independent affectional systems that can each in part compensate for deficiencies in the other."[70] Life is too precious to stake everything on one form of love.

The Moral of the Story: Surprise!

All animal fables end with a moral. To draw a moral about the innate
need for maternal love and care, psychologists such as Bowlby, Ainsworth,
and Spitz focused on Harlow's early results. Ignoring his later findings,
they continued to highlight the essential role of maternal love in early
infancy, almost exclusively when compared with other relationships.
Bowlby, for example, never cited Harlow's research on the role of peers.
As late as 1980, he discussed the importance of Harlow's findings as follows: "Furthermore, confidence that we are on the right track has been
enormously enhanced by experimental studies of rhesus monkeys. Harlow, for example, found that females which had been deprived of mothering during their own infancy grow up to be mothers which not only
neglect their infants but violently reject them."[71] Bowlby and Ainsworth

kept emphasizing the significance of the instinctual relationship between infant and mother. For them, as for many historical accounts, Harlow's work with monkeys revealed the power of biology and the deterministic role of experiences during infancy.

But all good stories end with a surprise. So I have kept the most intriguing results for the end. The infamous motherless rhesus mothers in Harlow's lab still had one last wondrous surprise in their experimental lives. After being cruel, brutal, and even lethal to their first newborns, most became "regular moms" when they gave birth to their second and third infants. This, Harlow recognized, was "a finding predicted by no member of the Primate Laboratory staff."[72] Could there be anything more surprising than a murderous mother? Perhaps only a brutal first-time mother who becomes a loving mother to her second infant. Yet this fact is rarely reported in accounts of Harlow's experiments. It is a different ending, and it changes the moral of the story: neither biology nor early experience determines an individual's destiny.

I do not see this as a story about the power of biology and the victorious return of instincts to explanations of behavior, as some commentators have argued. For example, historian Carl Degler has written that Harlow's work "constituted an important step in the rehabilitation of the concept of instinct." He sees this work as a major blow to the behaviorist emphasis on the role experience and training play in behavior. In a similar way, biologist Matt Ridley traces the fall of behaviorism to Harlow's early experiments, when the rhesus monkeys refused "to obey the theory that we mammals can be conditioned to prefer the feel of anything that gives us food" because, he states, "a preference for soft mothers is probably innate." Leaving aside the much larger historical question of the fall of behaviorism and its relation, if any, to Harlow's work, I don't think Harlow's results support the existence of instinctive behaviors as programmed actions that cannot be modified by the environment, because lack of social interactions resulted in a breakdown of the two behaviors most commonly taken to be natural: sex and maternal care. Although Harlow was sometimes ambiguous about his views on instincts, I have shown elsewhere that he did not support Lorenz's conception of instincts or the separation between innate and learned components of behavior. In my view, his experiments cannot be interpreted as supporting the triumph of instincts, as Degler and Ridley have claimed, or the instinctual basis of the mother-infant dyad, as Lorenz and Bowlby defended it. Furthermore, over time Harlow rejected the view that mother love is necessary for the emotional and social development of infant rhesus monkeys, while emphasizing the importance of peers and social stimulation.[73]

Neither did Harlow's experiments support the view that early experiences determine adult performance, as Lorenz and Bowlby believed. Monkeys socially crippled in infancy were later able to become "functioning" members of primate society. The motherless mothers who turned into adequate mothers after being socialized by their firstborn, even when they could not mother them adequately, showed this point in startling clarity. Neither biology nor the environment exerted a determinant influence over these resilient rhesus monkeys.

Furthermore, I also don't interpret this historical episode as an example of social determinism in which a hegemonic social consensus dictated the actions of experimental subjects and the scientific and social interpretation of those actions. First, experimental science did not exert a deterministic hold over the bodies of rhesus monkeys. In her analysis of Harlow's work, Haraway focused on "the boundary of translation and traffic between the laboratory and other areas of 1950–1975, U.S. middle class, white culture," but she examined the traffic only from society to laboratory, not from the laboratory to society. She concluded that the "primate body is a discursive construct and therefore a literal reality, not the other way around."[74] In her account, the monkeys were passively constructed through laboratory practices that conveyed social mores. Yet my analysis shows that monkeys did not always comply with expectations. Contrary to Haraway's reading, the monkeys did not simply model in the laboratory those behaviors researchers may have expected from them. More than once, the monkeys "surprised" the experimenters. In addition, society did not bring to the lab a monolithic discourse, as numerous members of the wider society, from journalists to politicians to mothers to women's clubs, became active in discussing the validity of Harlow's extrapolations from rhesus infants to human infants. The experiments were variously interpreted, sometimes as supporting the natural role of women as mothers, other times as destabilizing established gender roles in parenting.

Conclusion

Research on the nature of love turned Harlow into one of the most famous psychologists of the twentieth century. He presented his first experiments on affectional systems in rhesus monkeys in his 1958 presiden tial address to the American Psychological Association. Barely a decade later, in 1967, he received the National Medal of Science from President Lyndon Johnson, the Gold Medal Award of the American Psychological Foundation in 1973, and a mention in the Nobel Prize given to Konrad Lorenz, Niko Tinbergen, and Karl von Frisch that same year. In 1975 he was awarded the highest honor in psychiatry, the Kittay International Scientific Foundation Award.

Yet Harlow's work is often simplified, and the nuances of his positions are lost and misunderstood when they are presented as supporting the views of Konrad Lorenz and John Bowlby.

This chapter has followed Harlow's experimental work on the role of mother love and the development of affectional systems, his changing interpretations of the results, and the reception of his conclusions in the wider society from the late 1950s to the late 1960s. My examination does not support those accounts that see Harlow's work as providing experimental confirmation for Lorenz's views about instincts, or for Bowlby's views about the instinctual need for mother love. At first Harlow supported Bowlby's view about the key role of the mother in infant development, but later he departed from this position to emphasize the role of peers. In addition, Harlow did not think Lorenz's views about imprinting and instincts could be applied to the pliable development of affectional systems in monkeys or in humans.

Looking at Harlow's experiments over time reveals how experimenters, experiments, and rhesus monkeys all influenced each other in a process that underscores the open-ended character of science, biology, development, and love. By describing the twists and turns in Harlow's experiments over many years rather than focusing on one particular experiment or paper, we can appreciate the complexity of experimental results as well as the active role of experimental subjects, scientists, and various social actors in interpreting and appropriating those results in diverse ways. The sociohistorical setting did not determine a specific interpretation of the significance of his experiments for humans. Harlow's extrapolations from rhesus monkeys to humans proved rather controversial, and different audiences drew different implications. When his experimental subjects refused to behave according to the standard theories of the day, Harlow changed his interpretation of primate developmental needs.

The monkeys also revealed their own plasticity and resilience, belying contemporary expectations about the determinant power of biology and early childhood experiences. In a vision of child development that emphasizes inborn instincts and critical periods in infancy, there is no place for monstrous infanticidal mothers who later turn into loving caretakers. Mothers who turn into monsters fall outside the natural order; they simply become unnatural. As unfeeling entities they become more like machines, and in some small way this reestablishes a comforting order, for it allows us to separate the natural from the unnatural, the feeling from the unfeeling, the subject from the object. But by transgressing the boundary from the unfeeling to the feeling, monsters that turn into mothers are more unsettling to our simplistic notions of the natural and the unnatural. They reveal the complexity of nature, its permeability and flexibility. The lack of developmental determinism is the real monstrosity of nature, the real monster for science and for our hopes of controlling behavior and emotions.

PART 3

Naturalizing Nurture

CHAPTER SEVEN

The Nature of Love

*Mary Ainsworth's Observational and
Experimental Work*

We are here concerned with nothing less than the nature of love and its origins in the attachment of a baby to his mother. —Mary Ainsworth, *Infancy in Uganda* (1967)

Introduction

From the mid-1950s on, Bowlby appealed to Lorenz's studies of animal
behavior and to Harlow's work with rhesus monkeys raised with artificial mother substitutes to support his views about the instinctual nature
of the mother-child relationship. But by the late 1960s, as we saw in chapter 4, few animal researchers accepted Lorenz's conception of instincts.
And as we saw in chapter 6, Harlow's results during the 1960s showing the
importance of peers and play for the normal development of rhesus monkeys contradicted Bowlby's position. Aware that his unification of psychoanalysis and ethology required empirical support from studies on human
children and their mothers, Bowlby increasingly turned to the work of
psychologist Mary Ainsworth (1913–1999), whom many nowadays regard
as the cofounder of the ethological theory of attachment.[1]

From a position as assistant professor of psychology at the University of Toronto, in 1950 Ainsworth joined Bowlby's group in London
and started collaborating on a book with him and James Robertson. In
1954 she moved to East Africa and carried out research on infant care in
Uganda. After moving to Baltimore in 1955, in the early 1960s she contin-

ued observational work on mother-child relationships. Using an experimental design she called the "strange situation," she also observed children's reactions when their mothers left them alone with a stranger in a room at her laboratory. She claimed that the way children react to their mothers' departure and return reveals the strength of their attachment to them and the quality of the maternal care they receive. In her 1967 book *Infancy in Uganda*, Ainsworth put forward her work as confirmation of Bowlby's ethological theory of attachment behavior. In 1969 Bowlby published *Attachment*, in which he presented Ainsworth's research as providing independent empirical support for his theoretical views.[2]

In this chapter I examine the development of Ainsworth's views about attachment in the context of her relationship with Bowlby. Attachment researchers and many historical accounts present her contribution as the observational and experimental work that proved the validity of Bowlby's ethological theory of attachment behavior. They also often portray the professional relationship between the two as an ideal partnership, an image that Bowlby and Ainsworth also cultivated. However, my analysis of their correspondence reveals that their relationship was complex. Ainsworth was subservient to Bowlby, a position she tried to overcome over the years. In the latter part of her career, when her experimental studies helped support Bowlby's views, the relationship evolved into one of mutual benefit and support. I show Ainsworth's evolution from assistant to defender to independent researcher as she attempted to interpret her results within Bowlby's theoretical framework.

Here I also analyze the assumptions behind Ainsworth's work and the relation between her data and Bowlby's theory. I argue that her work was a key step in the reifying and biologizing of attachment. Her claims about the uniformity and universality of attachment behavior in children from different cultures—Uganda and the United States—shored up Bowlby's claims about the instinctive character of attachment. But my analysis indicates that neither her observational research nor her experimental work in the strange situation provided sufficient support for Bowlby's views about the biological basis of attachment behavior.

Mary Ainsworth: From Assistant to Defender

Born in Glendale, Ohio, in 1913, Mary Salter (later Ainsworth) grew up in Canada and studied psychology at the University of Toronto, where she earned a BA in 1935 and a PhD in 1939. In some of her early work, she studied security in infancy under the mentorship of William E. Blatz. For Blatz, children need to feel secure with their parents in order to use them as a base from which to explore and learn. In her thesis, "An Evaluation of Adjustment Based upon the Concept of Security," Ainsworth developed new scales to evaluate security and insecurity of young adults in their relationships. During World War II, she served with the Canadian Women's Army Corps, and in 1946 she became an assistant professor of psychology at the University of Toronto.[3]

After marrying L. H. Ainsworth, a graduate student in psychology, Mary Salter took a leave of absence from her academic position and accompanied him to London in 1950. She obtained a research position in Bowlby's group at the Tavistock Clinic. Here she met James Robertson and other researchers on children's issues and became more familiar with psychoanalysis and ethology. Bowlby, Robertson, and Ainsworth started collaborating on a book project under the name "Current Studies 1" (CS1), about how children react when separated from their mothers. The text would review the existing literature, critically analyze the observational studies, and provide a theoretical framework for explaining the effects of separation.

In 1954 Ainsworth moved with her husband to Uganda, where she continued work on CS1 and also carried out a study of infant behavior in relation to mothering practices among the Ganda people. Her original goal was to study infants' response to separation that did not involve deprivation as well. Most of the studies of separation up to that point had centered on children who not only were separated from their mothers but often were also sick and apart from other family members and friends. As several critics had pointed out, most probably these children's well-being and their response to separation from their mothers were influenced by those factors. Ainsworth thought the Ganda provided a perfect "natural experiment" to assess the effect of separation without complicating factors, because traditionally babies were separated from the mothers at weaning and sent to live with their grandmothers.[4] But her hopes did not materialize, because that pattern of child rearing was no longer practiced among the Ganda people when she arrived in Africa.

Ainsworth then designed a study of the lives of twenty-eight babies between one and eighty weeks old. She reported three main objectives: first, to examine the customs of infant care in Uganda; second, to analyze the development of interpersonal relations and attachment; and third, to

study the effect that variations in infant care practices had on development and attachment. As in other research about "interpersonal relations" in childhood, her study centered exclusively on the relationship
between children and their mothers. In exchange for helping them get
medical services for the children, Ainsworth was allowed to observe the
interactions between twenty-six mothers and their twenty-eight babies
during home visits (there were two sets of twins in the sample). She also
obtained information about the mothers' practices by interviewing them
with the help of an interpreter.

Ainsworth collected data, but she did not write up her results or her
interpretation of them during her stay in Africa. In 1955 she again moved
with her husband, to Baltimore. She was appointed associate professor at
Johns Hopkins University in 1958 and full professor in 1963. After taking
care of personal affairs and writing some other projects, she returned to
her work on attachment. Ainsworth continued what she called a "naturalistic" study in a longitudinal research project that became known as the
Baltimore Study. With some students, Ainsworth examined the relationships between twenty-eight babies and their mothers by observing them
during eighteen home visits to each pair, from childbirth through the first
fifty-four weeks of the infant's life.[5]

Despite doing research in Uganda and now in Baltimore, Ainsworth
published little of her work in this area during the 1950s and early 1960s.
In part, moving to different continents disrupted her writing. In Baltimore
she had to establish herself in a new academic home and in a new city
while also dealing with difficulties in her marriage. In part, she had devoted a lot of time to the book planned with Bowlby and Robertson.

In 1954 Ainsworth presented Bowlby's research group with a quite advanced manuscript of CS1. This ambitious project aimed to include over
thirty chapters. Ainsworth's draft was rather polished. The review of the literature on maternal deprivation and separation was an exhaustive, highly
systematic analysis of the available work to date, including tabulation of
data, comparison of hypotheses, and assessment of the conclusions. However, after reading Ainsworth's manuscript, Robertson and other members of the research group raised several objections. Some had to do with
her presentation of the material, but others were more substantive. They
focused on how much could be generalized from the detailed but limited
number of cases (about fifty) she put forward as evidence for her larger
generalizations. After extensive discussion, the group, including Bowlby
and Robertson, decided the project should be reorganized. The new

vision for the book would omit a lot of the material Ainsworth had written. Bowlby wrote to inform Ainsworth about these changes and invited her to come to London and meet with him and Robertson in April 1955.[6]

Eventually, and despite her protests, Ainsworth was left out of the CSI project. Bowlby and Robertson continued to work on the book, which they envisioned as the major presentation of their jointly developed views about the effects of separation.[7]

Although Ainsworth probably was deeply hurt by this decision, she continued her relationship with Bowlby and Robertson. Bowlby visited her in Baltimore in early 1958, when he held a fellowship at the Center for Advanced Study in the Behavioral Sciences at Stanford. He hoped that her work in Uganda would provide empirical support for his views. No longer the coauthor of Bowlby's planned major book, Ainsworth struggled to find a niche in relation to him intellectually and personally. During the early 1960s, her main publications were a defense of Bowlby's views, which were under heavy criticism.

In 1961, psychologist Lawrence Casler from Brooklyn College published an extensive review of work on maternal deprivation in which he aimed to assess two issues: first, "Is the concept of mother love useful in explaining the effects of maternal deprivation?" and second, "If so, which aspects of mother love are the operative ones?" Casler took as his starting point Bowlby's WHO report, for this was still considered the most influential summary of studies of maternal deprivation. After reviewing all forty-five studies Bowlby cited to support his views, Casler noted that many of them referred to the same group of children. For example, several papers by William Goldfarb were descriptions of the same group of fifteen institutionalized children and fifteen controls. Some other studies dealt with only two cases of children hospitalized for serious conditions. In yet another study the authors had noted that the ill effects of institutionalization were reversible, contrary to what Bowlby claimed. For criticisms of Margaret Ribble and René Spitz, Casler referred mainly to other reviews, especially Samuel Pinneau's work. He noted that the other studies Bowlby cited failed to take into account many critical variables, such as the age when the separation occurred, the nature of the institution where the child was raised, and the reason for the separation (for example, whether the child was sick or had been unwanted). Casler concluded that "without exception" the works Bowlby used were "neither conclusive nor particularly instructive."[8]

In addition, Casler pointed out a variety of methodological problems with the studies Bowlby relied on. These problems concerned the selection and size of the samples and the validity of the criteria used to measure pathological deviations. Most studies ignored age, did not handle the data in a statistically meaningful fashion, and neglected the difference between deprivation of "love" and deprivation of those forms of perceptual and sensory stimulation that "frequently accompany expressions of affection." "None of Bowlby's references offers satisfactory evidence that maternal deprivation is harmful for the young infant," Casler concluded. He was not alone in his low opinion of studies of maternal deprivation. In another extensive review of the literature, psychologist Neil O'Connor reached the same damning conclusion. Gradually, other voices joined the critical chorus.[9]

Bowlby's work became so controversial that in 1962 the WHO published *Deprivation of Maternal Care: A Reassessment of Its Effects,* barely a decade after Bowlby's 1951 report *Maternal Care and Mental Health.*[10] All the papers, with one exception, criticized Bowlby's views because of the lack of supporting evidence. After an examination of recent studies, Dane G. Prugh and Robert G. Harlow concluded that one could not establish either a necessary or a sufficient relation between maternal separation and mental illness. R. G. Andry argued that, contrary to Bowlby, more recent investigations tended to show that maternal deprivation was not an important factor in most cases of deviant behavior. He also pointed out the need to study the role of fathers before concluding that the mother had a determinant role in child development. Serge Lebovici, secretary general of the Institute of Psychoanalysis in Paris, took a more appreciative look at Bowlby's work, but even he concluded it would be "highly dangerous to attribute the overwhelming majority of emotional and mental disorders in adolescents and adults" to maternal deprivation. He emphasized the need to pay attention to the diversity of stimuli that children need.[11]

Another major criticism focused on Bowlby's methodology. In an impressively detailed analysis of his work, the English social scientist Barbara Wootton listed various ways it failed to meet contemporary scientific standards. For one thing, she noted, Bowlby "paid too little attention to the findings of studies which run counter to his theory." It was also difficult to interpret the evidence he cited, for several reasons: the samples were small; most studies were done with children from institutions or hospitals; the studies did not pay attention to other factors such as the children's diet and backgrounds; most studies did not have control groups; and it was not easy to distinguish between separation and deprivation.

In addition, no evidence existed to support Bowlby's view that the damaging effects of separation were irreversible. Wootton concluded: "Whatever the future may show, reference in the present state of knowledge to the 'permanent,' 'irreversible' or 'irreparable' damage due to separation is reckless and unjustified."[12]

Margaret Mead, who was familiar with Bowlby's writings and knew him personally through the WHO and other scientific meetings, argued that the anthropological evidence did not support his claims about the universality of the "monotropic" mother-child attachment. Bowlby's emphasis on monotropy, the need for a single mother figure, rested on the assumption that there is a biological need for continuity in the mother-child relationship. As Mead put it, this implied "a pair relationship which cannot be safely distributed among several figures," and it supposed that "all attempts to diffuse or divide it and all interruptions are necessarily harmful in character, emotionally damaging, if not completely lethal." However, she noted that her own studies on Samoa did not support this thesis, since children who were cared for by different people became well-adjusted adults.[13]

But one could go even further, Mead noted, and ask, as Konrad Lorenz did, whether some cultural practices are lethal for human society, if not for the individual child. The question was "whether the cultivation of more exclusive and more intense parent-child relationships is not a precondition of the kind of character structure which is necessary to maintain and develop our kind of civilization."[14] In her phrasing, Mead encompassed two issues: whether the mother-infant dyad is necessary for the survival of the human species, and whether it is necessary for the continuity of Western civilization.

For Mead, Bowlby was not justified in asserting that a monotropic relationship between child and mother is a universal and uniform part of human nature. Studies of other cultures, including the Israeli kibbutz, the Hutterites, and extended families in China and India, had shown that a large number of nurturing figures besides the mother could also provide the security children need for healthy emotional development. She added that by giving the mores of his own culture the status of universal behavior and positing a biological underpinning, Bowlby had committed the sin of "reification." He had taken "a set of ethnocentric observations on our own society, combined with assumptions of biological requirements which are incompatible with Homo sapiens," and turned them into "a set of universals."[15]

In the 1962 WHO volume Mary Ainsworth wrote the only article supporting Bowlby's position, and her contribution was designed for just that
role. Rather than writing a defense of his position himself, Bowlby left
this task to his former associate. In correspondence, the volume's editor
said he wanted Bowlby's views to be "fairly represented," and he com-
mended Ainsworth for her presentation of "the deprivation case." A loyal
defender, she sent him reports of her progress while she prepared a response to his critics. Ainsworth met with Bowlby and discussed her draft
of the paper, which he also shared with Robertson. Although both were
"laudatory" and considered the paper "an important contribution," they
suggested changes that Ainsworth incorporated into the paper as she re-
wrote it.[16]

In "The Effects of Maternal Deprivation: A Review of Findings and
Controversy in the Context of Research Strategy," Ainsworth's main tac-
tic was to argue that Bowlby's critics applied unfair epistemological standards. She claimed that his detractors set the benchmark too high because
"the experimental method, the backbone of laboratory research, has limited applicability to the study of maternal deprivation."[17] In this way she
suggested that Bowlby did not have experimental evidence because nobody could have it. Yet she did not address the main criticisms regarding
his lack of sufficient observational evidence for his position and his disre-
gard for observational evidence contrary to his views.

Ainsworth paid special attention to the paper by Margaret Mead. According to Ainsworth, Mead mistakenly assumed that Bowlby "sponsors
an exclusive mother-child pair as the ideal." However, she added, Bowlby
had "argued for the desirability of a major mother-figure, not necessarily the biological mother, whose care is supplemented by other figures,
including a father-figure." Yet Ainsworth also said that "dispersion of
maternal care" was "not likely to be the norm in any primitive society."
Furthermore, in her view it was "entirely likely that the infant himself
is innately monotropic." As a consequence, she pointed out, "a situation
(whether brought about by an 'experimental society' or through some individual variation in a traditional society) which impedes monotropic attachment will distort the normal course of development."[18]

What a strange combination of arguments! Ainsworth first declared
that Bowlby did not posit monotropy with the biological mother as the
ideal, but then she proposed that monotropy was probably the norm in
primitive societies, that monotropy was probably innate, and that it would
be dangerous to disturb this innate pattern. If the first argument was correct, why would Ainsworth go on to make the other three? In any case,

she did not provide evidence for them. Moreover, regarding her first claim, we have already seen that Bowlby did support monotropy and attachment to the biological mother as the ideal. Indeed, this claim is essential to his ethological theory of attachment, which asserts that maternal responses are instinctive. As we saw earlier, Bowlby claimed that children need maternal love and care "365 days a year." But why would nature design women to care instinctively "365 days a year" for children who are not their own?[19]

The discussion in the WHO report made it clear that Bowlby needed to provide empirical support for his views about the mother-infant dyad. As Ainsworth began to offer it, Bowlby became increasingly eager for her support. In these early years, their correspondence reveals Ainsworth's dependent situation. Bowlby saw himself as the theoretician of attachment, and Ainsworth often accepted her secondary status.

The story of her joint book project with Bowlby and Robertson provides a clear example of her subordinate position. After she was left out of the project, Bowlby and Robertson continued work on the manuscript, now with the tentative title *Protest, Despair, and Detachment*. Originally they planned to combine Robertson's observations and their jointly developed ideas about the effects of separation with Bowlby's views about the biological basis of attachment. Bowlby published independently the central theoretical chapters, but meant to include them as chapters of the book. That book would have been the major presentation of the attachment framework. Eventually, however, they finished for joint publication a manuscript of more limited scope. In 1964, Bowlby sent Ainsworth the new version of the book for her comments. She read it carefully and sent Bowlby incisive criticisms. She also wrote to Bowlby:

> I thank you for your acknowledgment of my contribution in your preface. I still wish, however, that you and Jimmy could see your way clear to include me as a third author. I read the manuscript with so much feeling of recognition. I know that there have been substantial changes since the first draft which I wrote—and I know that some of what I would like to see changed (especially in regard to changes in terminology dealt with in my point (2)) is a carry-over from that early draft—but nevertheless I read the manuscript feeling that what I was reading was a revision of *my* draft. I do not feel as strongly about this as I did when I first protested your decision to omit my name. Then I felt that I had devoted two years of hard work to a project that was not represented in my list of publications. Now that promotion has been achieved, and there is no crucial importance attached to my list of publications, I feel less strongly.[20]

In spite of Ainsworth's new plea, Bowlby and Robertson refused to include her as an author. Although this would have been the only publication she would have had to show for a long period of extensive work on attachment, they did not consider that her work merited coauthorship. They discussed this in several letters, but to no avail.[21]

Ainsworth's early draft was never published, and neither was Bowlby and Robertson's book. Many of the observations and the ideas discussed in those manuscripts found their way, in modified form, into Bowlby's later books. The history of the unpublished book manuscripts, however, provides valuable insight into the factors that play a role in determining who gets authorship and what type of work deserves scientific recognition. In this case, many factors became intertwined in the complex dynamics of interdisciplinary group research: the researchers' gender, their social class and status, the unequal authority of their disciplines and diverse traditions and goals, as well as their very different personalities. In their discussions, Ainsworth pushed for systematic statistical analysis and presentation of the data. Bowlby was interested in generalizations that would allow him to develop a new theory about child development. And Robertson continually asked them to present his early observations not as proof of general claims, but rather as illustrations of some preliminary views about the effects of separation on children in specific circumstances. Over the years, Bowlby and Ainsworth told Robertson he was being too modest about his data. In addition to personality differences, the complicated relations between the experimenter, the theoretician, and the social worker were also affected not only by their different status in the field, but also by the different status of theory, experiments, and observations in scientific research.[22]

For Ainsworth, her attempt to find her own personal and professional ground while maintaining a good relationship with Bowlby became an ongoing struggle. Over the 1960s, Bowlby developed his ethological theory of attachment behavior. After reading Thomas Kuhn's influential account of the history of science as a series of paradigm shifts, he conceived this theory as a "new paradigm" in child development. Ainsworth strove to interpret her observations of children within that paradigm, but she did not want to present her work simply as a set of observations that confirmed what Bowlby had already established. She eventually won the battle for recognition by convincing Bowlby that publishing her work independently would benefit him as much as her:

> I feel, and have felt for a long time, that it is to our mutual advantage for me to present my ideas as _my_ ideas rather than as mere echoes of yours. This is dif-

ficult because our thinking is so very much in track. I neither like to deprive myself of your formulations (which are much better thought through than my own) or of the backing which quoting you brings, nor do I like to present myself as a devoted disciple and no more. The latter does me an injustice—and also weakens the confirmation that I can provide for you. I think it has been of some value to you that I have been more or less independent, and, when relevant, you can quote my findings as findings independent of your own research endeavours.[23]

Ainsworth thus developed a successful strategy for gaining recognition for her own contribution without alienating Bowlby. She openly acknowledged his superiority as a theoretician. "Who am I to have a theory?" she wrote to him once.[24] But then she emphasized how much her own empirical work could help him. In this way, she hoped to escape the "master-disciple image which would detract from both of us," as she had written to Bowlby earlier.[25]

Today many attachment researchers present Bowlby and Ainsworth as coauthors of the ethological theory of attachment. Yet their extensive correspondence reveals the tension underlying the "happy partnership" that was the public face of their relationship. But in their writings we see this assertion only in their 1991 article "An Ethological Approach to Personality Development," which explicitly described the history of their research as a "happy partnership." However, as explained in an author's note, Bowlby died in September 1990. Ainsworth wrote the paper, and it is telling that this was most likely the first time she presented herself as coauthor of the theory of attachment.[26]

Bowlby and Ainsworth's personal and professional dynamics are important for understanding the evolution of her views and assessing their relation to Bowlby's framework. In her 1962 defense of Bowlby's views discussed earlier, Ainsworth asserted that the experimental method had limited applicability to the study of maternal deprivation. Yet a few years later, she presented her work as providing both the "natural" and the laboratory-controlled experimental data to support Bowlby's views on attachment.

Patterns of Behavior: From Uganda to Baltimore via London

In 1967 Ainsworth published *Infancy in Uganda: Infant Care and the Growth of Love,* in which she interpreted the results of her research in

Uganda and Baltimore. She had organized her notes and published parts of her African research earlier, but she did not advance a full interpretation of her data until 1967, more than a decade after she carried out her observations. In her book she noted how the results obtained in Baltimore shaped her views about Uganda and the other way around. In addition, by this time Ainsworth had read Bowlby's views about the ethological theory of attachment behavior and had discussed them with him frequently. She and Bowlby met several times in the United States, and she had also participated in some of the meetings he organized at the Tavistock Institute, where many researchers doing empirical work on the mother-infant relationship presented their results. Here she discussed some of her data from Uganda in 1961.[27] In her 1967 book she offered a unified interpretation of her studies as supporting Bowlby's theory.

Ainsworth divided her book into four parts: one dealing with methodology; another focusing on the Ganda's methods of infant care; the third containing several case studies with descriptions of individual babies and families; and a fourth, more theoretical chapter titled "Development of Attachment." Most of her views were based on her observational research with infants in Africa, but she also included her conclusions about the Baltimore study.

The group Ainsworth studied in Uganda was a varied lot in many respects. She observed fifteen males and thirteen females (including two sets of twins). The children lived in six villages near Kampala. They ranged, at the beginning of the study, from two days to eighty weeks old. Fourteen babies lived in monogamous households, and seven were part of polygamous households. Three babies had unmarried mothers; in three other cases the fathers were absent; and in one case the parents had separated. None of the babies lived alone with their mothers. Father surrogates were also common. Several of the families had adopted Western practices of child rearing, such as scheduled feeding. The mothers used books or the local clinic for information on child care and nutrition. In spite of all those differences, Ainsworth claimed that the settings were "so similar from one household to another that it provided a fairly standard situation in which we could observe individual differences in behavior."[28]

Ainsworth combined observing mothers with their infants and interviewing the women. She observed the interaction between each mother and her infant during scheduled afternoon visits, interviewing the mothers just before the observation period began to find out about daily events she could not observe during the visit. For the interviews she relied on

FIGURE 7.1. A Ganda mother and her children. From Mary Ainsworth, *Infancy in Uganda: Infant Care and the Growth of Love*, p. 203. © 1967 The Johns Hopkins University Press. Reprinted with permission of The Johns Hopkins University Press.

Mrs. Kibuka, her interpreter, to translate her questions and the mothers' responses. Ainsworth had planned to make a two-hour visit to each family every two weeks, but she ended up making briefer, more frequent visits. She reported the median number of contacts with each family as twenty-three.

To study the infants' behavior, Ainsworth observed their crying, smiling, vocalizing, following, visual-motor orientation, lifting arms in greeting, clinging, and exploring. During the interviews she obtained additional information about the household, birth, feeding, elimination, cleanliness, sleeping, motor development, vocalization, crying, distribution of mothering duties, interpersonal relations (responses of baby to people and behaviors), sexual behavior, aggression, punishment, dangers and concerns, play, health and medical care, weaning, and separation.[29]

To study mothering practices, Ainsworth first observed the quantity and mode of physical contact between mother and child. She noted that mothers carried their babies variously on their backs, hips, arms, or chests. She also noted the difficulty of assessing cuddling as an index of warmth in interaction, because "cuddling" could not be translated into Luganda. Second, she found out if mothers shared their mothering duties with other people. This was of great interest because the extent and exclusivity of "attachment to the mother" was at stake.[30]

Ainsworth first elaborated a list of patterns of behavior that served as her criteria for assessing whether an attachment has been formed. They related to three basic categories: one, the infant's ability to identify her mother; two, the infant's concern for her mother's whereabouts and departures, as well as different greeting responses; three, the child's behavior in the mother's presence, such as wanting to be close to her even with no threat of separation. Ainsworth maintained that the child's conduct in these three areas reflected the strength and security of attachment. Then, using these criteria, she divided the children in her sample into three groups: securely attached (sixteen children); insecurely attached (seven children); and nonattached (five children).

Ainsworth noted that most of the insecurely attached children were chronically malnourished or ill. The others were "rejected by [the] mother," or the mothers were "anxious and depressed."[31] But she also thought that their mothers' care practices contributed to their insecurity.

Since she observed those children only during a brief period, Ainsworth could not assess if the different types of attachment affected a child's future. However, she appealed to an alleged scientific consensus on the significance of attachment in infancy for the development of the adult personality. Thus she wrote:

> On the basis of all that we know about the effects on subsequent development of deprivation of maternal care in infancy and early childhood, it is an essential condition of subsequent satisfactory interpersonal relations for a child to have formed an attachment to someone—his mother or a substitute for her. On the basis of all of our clinical knowledge of the relationship between early parent-child relations and later intimacies, it seems desirable for a baby to have a secure attachment to his mother rather than an insecure one.

Taking this as a given in her own work, she then asked, "How then can infant-care practices foster the development of a secure attachment?"[32]

To answer that question, Ainsworth next analyzed the mothers' con-

duct to evaluate its impact on a child's attachment. Aware of the limitations of her study, including the small sample size and the lack of continuous and complete observations, Ainsworth noted that this could be
answered only "tentatively."[33] She separated the groups according to two
sets of variables: variables associated with feeding, and other variables
such as warmth of the mother and multiple caretakers.

Ainsworth found the analysis of feeding variables inconclusive. She
studied scheduled versus self-demand feeding as well as the impact of the
mother's milk supply. She thought the sample was too small for statistical tests to show anything significant. However, she concluded that the
mother's attitude toward breast-feeding did make a difference: of thirteen
women who said they enjoyed breast-feeding, twelve were mothers of securely attached children, and none were mothers of nonattached children.

With regard to the second set of variables, Ainsworth found the effect
of multiple caretakers irrelevant to her study. She noted that this had
been a point of contention between Bowlby and Margaret Mead. Mead
had disputed the presence of a single, continuous mother figure as a necessity for healthy emotional development. Ainsworth also stated that in
her study most mothers shared caretaking. It is not clear, then, why she
did not cite the children who were taken care of by several people and
who were securely attached as evidence in favor of Mead's views. Ainsworth simply did not address the issue.

Ainsworth focused on the warmth of the mother and, specifically,
on two characteristics she had found to be important in her study of
American babies in Baltimore: "the sensitivity of the mother in responding to the baby's signals and the amount and nature of the interaction
between mother and baby." She aimed to differentiate between routine
care and genuine, sensitive interaction. In her Uganda study she said she
also found another variable that correlated positively with attachment:
the mother's excellence as an informant.[34]

What was at stake for Ainsworth was the mother's interest in her
baby, her anxiety, and her other preoccupations. A mother who was at-
tentive to her baby received a high score. A mother interested in "other
topics of conversation" and "preoccupied with other thoughts or activities" received a low score. Mothers of nonattached infants were found to
be below the median as informants about their babies, whereas mothers
of securely attached infants had very high ratings. "This variable significantly differentiates between the three groups and hence seems related
to the development of attachment," she argued. Ainsworth conceded that

"it is conceivable that a mother might be uncommunicative in interview for reasons totally unrelated to her attitude toward the baby," but she believed this was not so in her sample. She did not explain why.[35]

To interpret the general variable "amount of time, care, and attention" from the Uganda study, Ainsworth turned to her ongoing study in Baltimore. There she reported that securely attached babies all enjoyed the following: frequent and sustained physical contact with their mothers; mothers able to soothe them effectively through physical contact; mothers sensitive to their signals and responsive to their demands; caretaking timed in harmony with the babies' rhythms; and mothers regulating the babies' environment, thus enabling predictability. According to Ainsworth, the typical Ganda mother did all these things.[36]

She found that three variables "related significantly to the development of security and attachment." These were the mother's attitude to breast-feeding, the amount of care she gave the baby, and her excellence as an informant. In Ainsworth's view, two of the significant variables from the Uganda study, mothers' excellence as informants and mothers' enjoyment of breast-feeding, corresponded to the "mutual delight" variable in the Baltimore study.[37]

Putting together the results from the Uganda and Baltimore studies, Ainsworth concluded that the development of an infant's security and attachment to mother was related to

> the sensitivity of the mother in responding to the baby's signals of need and distress and to his social signals, and the promptness and appropriateness of her response; the amount of interaction she has with him and the amount of pleasure both derive from it; the extent to which her interventions and responses come at the baby's timing rather than her own; the extent to which she is free from preoccupation with other activities, thoughts, anxieties, and griefs so that she can attend to the baby and respond fully to him; and finally and obviously, the extent to which she can satisfy his needs, including his nutritional needs.[38]

Having uncovered the factors influencing the development of attachment, Ainsworth turned to measuring attachment in the laboratory. In the early 1960s, in the context of her longitudinal study of infant care in Baltimore, she designed an experimental setup she called the strange situation, reminiscent of similar experimental setups used by psychologists Jean Arsenian and Harry Harlow. In the mid-1940s, Arsenian had examined the reaction of children from eleven to thirty months of age when they were alone or when they were accompanied by their mothers or another adult in a strange room full of toys and pictures. She concluded that their reaction depended on the unfamiliarity of the environment and the power of the person with them in that strange situation. As we saw in chapter 6, in the 1950s Harlow used "the strange room" to observe infant monkeys' reactions to stimuli in the presence or absence of their artificial surrogate mothers.[39]

In Ainsworth's setup, a mother, her infant, and a stranger interact in an eight-stage, twenty-minute experiment. The experiment begins when a mother and her infant enter a room in the laboratory. The mother sits down while the baby plays on the floor with some toys. After a few minutes, a stranger enters and also sits down, engaging the mother and the baby. The mother leaves the room for a few minutes, leaving the baby alone with the stranger, who is to offer comfort if the baby is distressed. After the mother returns the stranger leaves the room, and the baby goes back to playing on the floor. Then the mother leaves the room and, after a couple of minutes, the stranger reenters. The stranger interacts with the baby for another couple of minutes, then the mother reunites with her baby. A team of observers records the experiment, focusing on the baby's reactions.

In her original presentation of the experiment at the fourth CIBA Foundation meeting in 1965, Ainsworth and Barbara Wittig concluded that "this study suggests that stranger anxiety is in a one-to-one relationship to neither attachment nor separation anxiety." But the strange situation experiment became increasingly important in Ainsworth's views, and she soon started to see it as a diagnostic tool for attachment. In 1967 she told Bowlby, "I am gaining more and more confidence in this strange situation procedure as a one-shot way of assessing attachment." In developing the strange situation procedure, the focus shifted from the child's reaction to the stranger and the use of the mother as a base for exploration to the child's reaction when the mother returns. An infant who accepts the mother's leaving and is happy when she comes back is considered securely attached. An infant who cries and is upset when the mother returns or is unresponsive is considered insecurely attached. She concluded that the categories established in the strange situation are good indicators of the quality of the relationship between mother and child.[40]

Although Ainsworth did not put it this way, she was making the case for three major theses: that infants' behavior clearly reveals identifiable patterns; that those patterns indicate different types of attachment; and that the type of attachment is related to the mother's behavior and emotions.

FIGURE 7.2. Mary Ainsworth (*right*) with unidentified woman and child, 1973. Mary Ainsworth images collection (image 05747). Courtesy of the Rare Books and Manuscripts Department, The Sheridan Libraries, The Johns Hopkins University.

Yet Ainsworth's interpretation of her observations and, later, Bowlby's use of her work relied on a series of problematic displacements.

Assumptions and Displacements:
From Relation to Correlation to Causation

In following Ainsworth's interpretation of her data over the years, one can see several displacements that undermine her conclusions. Some concern the concepts she used, while others involve her claims about the relation between maternal care and attachment. Here I explore these briefly, after first noting some problems with her data.

How robust and reliable were Ainsworth's observations? She recognized that in Uganda it was difficult to obtain data about some variables. Since medical assistance and transportation to a clinic were a reward for participating in her study, it is likely that her sample was biased toward parents concerned about their children's health. This concern probably affected how communicative the mothers were as well. In addition, an interpreter translated both Ainsworth's questions and the mothers' answers. Ainsworth herself noted that some concepts were not easy to translate. It is thus difficult to evaluate how much her perception of a mother's sensitivity toward her child was influenced by the translator's mediation and the ambiguities in both languages.

Ainsworth's data about maternal care in Baltimore homes may not have been completely reliable either. Various students who worked as her research assistants observed mothers with their children in their homes. Working from a list of behaviors to observe, these students made some notes and wrote their reports after the home visits. In at least one case, the assistant did not write her reports until many months later. Ainsworth herself realized that the reliability of those reports was compromised, as she wrote to Bowlby:

> For two years, I set myself to believing that I had found a good assistant, and to ignoring the obvious deficiencies in her performance. Suddenly, there was a moment of truth. I found that she had written up fewer than half of the visits that she had made to the babies in our sample. She has been catching up ever since, and will probably not finish catching up until the end of September when the grant runs out.

Ainsworth added that the student had taken very good notes and the reports written months after the home visits still had "freshness and vividness (and, as far as I can judge, reliability)."[41] But it is hard to say how much of this was wishful thinking or a way of reassuring Bowlby, eagerly waiting to include her results in his book on attachment.

Another word of caution is necessary regarding the relation between her data on infant care in Uganda and Bowlby's theory, because Ainsworth interpreted those observations about ten years after collecting them. Although it is always presented as work done in the ethological tradition, and often presented as providing independent confirmation of Bowlby's theory, I think it would be fair to say that her later work, interests, and collaboration with Bowlby influenced her interpretation of her data on Uganda. All observational work is at some level mediated by theory, and this alone does not invalidate her work. Nevertheless, it would be incorrect to say that her observations provided *independent* confirmation of Bowlby's theory; instead, she used his theory to interpret her observations.[42]

In addition, it is unclear how Ainsworth interpreted the statistical significance of some data. For example, of the thirteen mothers who said they enjoyed breast-feeding, twelve were mothers of securely attached children, and none were mothers of nonattached children. Ainsworth concluded: "It is thus obvious that this variable significantly differentiates the groups."[43] But there were also four securely attached children whose mothers did not enjoy breast-feeding. Ainsworth did not account for this. To give another example: in her Uganda study, she found that in a number of cases mothering was carried out by several people, so she discounted this as a variable. But given the importance of monotropy in Bowlby's theory, why didn't she cite the fact that most of the children in her study were securely attached as evidence against monotropy?

Perhaps aware of some of these issues, Ainsworth emphasized the tentative nature of her conclusions. She couched them as conjectures, and she appealed to Harlow's work on rhesus monkeys and to Bowlby's theory: "But let us not carry the argument further on the basis of the Ganda evidence. Harlow has much more telling evidence from his study of infant monkeys, and Bowlby has argued the theoretical points more cogently."[44]

Despite recognizing the shortcomings and limitations of her data, Ainsworth was quite ambitious about their significance: "We are here concerned with nothing less than the nature of love and its origins in the attachment of a baby to his mother." Love, no less, and its nature. This is what was at stake for Ainsworth in studies of attachment: "What has emerged [from the past eleven years of research and interpretation] is a new way of viewing the origins and early growth of first love—the attachment of a baby to his mother."[45] On this point she followed the psychoanalytic tradition of seeing the mother-child relationship as the source of a child's general capacity for love. Like many psychoanalysts, including Bowlby, Ainsworth made mother love the main source of one's ability to love, the prime mover of one's emotional life. Unlike psychoanalysts, though, she and Bowlby saw the infant's attachment to the mother not as a derived feeling, but as a primary one.

But what exactly is attachment? Ainsworth said that "attachment originates in a few specific patterns of behavior, some of which are manifest at birth and some of which develop shortly afterward." In a long passage describing the nature of attachment, it is evident that she struggled to come up with a definition of the concept and to specify its referent:

> Attachment is not present at birth, however; it emerges gradually through a course of development, and it is perhaps a matter of more or less arbitrary definition to identify the point at which it could be said to have finally emerged.

Attachment is manifested through these patterns of behavior but the patterns do not themselves constitute the attachment. Attachment is internal. We can conceive of attachment as somehow being built into the nervous system, in the course of and as a result of the infant's experience of his transactions with his mother and with other people. This internalized something that we call attachment has aspects of feelings, memories, wishes, expectancies, and intentions, all of which constitute an inner program acquired through experience and somehow built into a flexible yet retentive inner mechanism (which we identify with central nervous system functions) which serves as a kind of filter for the reception and interpretation of interpersonal experience and as a kind of template shaping the nature of outwardly observable response.[46]

Here Ainsworth made an important displacement from behavior to feelings.

Ainsworth recognized that behavior does not automatically reveal inner feelings and emotions. This is clear in other passages where she justified her framework. She maintained that there were two approaches to the "inner life of an infant." One was psychoanalysis, which for her relied on the adult's recollections and fantasies to get at the infant's inner life. "Another approach, and the one I have adopted," she continued, was "to observe the infant's behavior, which is undoubtedly related to his inner experience, although not a transparent communication thereof." When explaining her adoption of the ethological concept of instinct, she also noted that the main difference from psychoanalysis centered on the fact that psychoanalytic drives were "internal and are not directly observable," whereas ethological instincts refer to behavior, which is "external and observable."[47] Though she asserted that behavior and feelings are different, Ainsworth often moved from one to the other in her own work.

Concerning attachment and maternal love, she also moved problematically from relation to correlation to causation. Cautious at first, Ainsworth talked only of relation. She also recognized the small sample sizes of her observational studies in Uganda and Baltimore. Nevertheless, she then claimed that her sample was "heterogeneous enough that hypotheses of cause and effect relationships between infant care and 'outcome' can be formulated." The search for cause and effect, in her words, "has the strength of exploiting to the utmost the relationship between an antecedent condition which appears to be the cause and a consequent outcome which appears to be the effect." Still, she acknowledged that "not all antecedent conditions are in a cause and effect relationship to events that follow them."[48] In her first presentation of the strange situation, she also pointed out the limitations of the conclusions: "Despite the satisfactory level of statistical significance of the group trends, the findings of this study cannot be generalized beyond white, middle-class American one-year-olds who have been reared much as these babies were reared."[49]

Nevertheless, Ainsworth posited the existence of a causal relation between different mothering practices and maternal feelings and the degree of children's attachment to their mothers. She never explained how she was justified in going from correlation to causation. She had done some studies of children and mothers in Uganda and Baltimore, observing them at home and carrying out experiments in the strange situation setup. At best these observational studies could show that certain patterns of maternal care were correlated with certain child behaviors. But that correlation is not sufficient evidence to argue for a causal relation unless all other factors that could influence the child's behavior could be controlled for, something Ainsworth never attempted.

It is even problematic to use the data from the strange situation experiments to argue for a correlation, since they relate only to one behavior—how children react when their mothers leave them alone with strangers. It is already a stretch to claim that a single behavior reveals attachment, unless attachment is defined as what is measured in the strange situation, which would make the reasoning circular and faulty. Moreover, to claim that the attachment behavior is a reliable indicator of maternal care and love requires a much greater leap. This is not justified unless one assumes that maternal care is the only factor, or at least the main factor, that influences how an infant reacts to strange people and situations. But what about the child's physical well-being, personality, and habituation to strangers? It surely must make a difference whether a child is raised alone or with siblings, in a nuclear or extended family, is often left with neighbors, friends, or sitters, and so on. A variety of factors will play a role in an infant's reaction to unknown people in an unknown setting. Thus, from a child's reaction to a mother's absence, the psychologist cannot infer the quality of maternal care and love.

Furthermore, she supposed that the child's behavior reveals not only the quantity and quality of maternal care the child receives but also the mother's feelings. In spite of her emphasis on observing behavior, for her what really mattered to ensure the development of a healthy attachment between mother and children was "the sympathetic" love that a mother "intuitively" feels for her child, and her "unconscious feelings." Thus Ainsworth, much like Bowlby, said that mothers who provided the correct amount of physical contact "because they believed the babies needed it, although they themselves did not enjoy it, ended up with improperly attached babies."[50] Somehow, by observing the conduct of children, attachment researchers claimed to uncover the feelings underlying a mother's behavior—her hidden and unconscious emotions.

Like so many authors writing about these issues since the 1950s, neither Bowlby nor Ainsworth explained precisely what was meant by a mother's "ability to time her interventions in harmony with the rhythms" of a child. Nor did they explain how a baby could detect the mother's feelings of delight. Just as important, how could the scientist measure the "mutual delight" that mother and infant felt in each other's company? Can a scientist assess this by visiting a mother only twice a month? And what do we know for certain about the "emotional rhythms" of infants anyway? Ainsworth noted the limitations of her data and acknowledged the difficulty of deducing emotions from behavior. However, she and Bowlby thought they could discern from children's behavior what went on in their mothers' hearts.

In the 1982 second edition of his book *Attachment*, Bowlby also presented Ainsworth's work as providing "extensive evidence of a strong correlation between the pattern of attachment observed in an infant or older child and the pattern of mothering he is receiving at the time." He further claimed there were "also clear indications that the pattern of attachment a child is showing towards his mother figure is to a high degree the consequence of the pattern of mothering he is receiving."[51] So not only was there a correlation, but maternal care was *the cause* of the child's behavior.

Ainsworth also moved from behavior to instincts, a displacement that deserves separate consideration.

The Biological Foundations of Attachment

When discussing the variables influencing attachment in her study of babies in Uganda, Ainsworth acknowledged that "in the present study it is quite impossible to differentiate genetic, prenatal, and perinatal influences from environmental influences since only one baby was observed from birth onward, and even in this case observations were too widely spaced to be helpful." She argued against "students of infant behaviour" who thought that development is merely "a maturational process of unfolding of genetic potential" and for whom infant care and child rearing were irrelevant. She argued equally against psychologists "chiefly concerned with animal learning," who "equate development with learning" and emphasize only "environmental contingencies."[52] Here she seems to support an interactionist view of environmental and biological factors. Furthermore, she noted the difficulty of doing so and, in particular, the inability of her study to separate genetic and environmental influences.

Nevertheless, Ainsworth gave biology a determinant role because she saw attachment as a universal preprogrammed schema. She first noted that "attachment does not develop willy-nilly according to some inner, genetic, regulating mechanism, but rather is influenced by conditions in the baby's environment." But then she added that genetics provides a specific structure to the developmental process:

> This does not imply that inner, genetic regulators are negligible. On the contrary, the orderly sequence in the emergence of attachment patterns—as well as the fact that there seem to be striking similarities between this sequence among the Ganda and the sequence observable in our society—suggests that the development of attachment is correlated with other processes of development and that all these processes are in part *determined by genetic factors characteristic of the human species.*[53]

Ainsworth presented herself as following "the ethologists (and Bowlby) in talking of instinctive behavior." She thus supported the view that

> the child is born with a behavioral repertoire which includes responses already biased toward mediating social interaction—and that this is part of his built-in, species-specific equipment. In addition to actions which have a primarily social function, his repertoire includes some which have potential multiple functions and can be turned to social ends. Thus the potential for attachment is part of the equipment of the newborn, although obviously attachment itself is only gradually acquired in the course of development—acquired in the course of interaction with his environment and the feedback he experiences as a consequence of his actions.[54]

At times Ainsworth said she was not using the discredited traditional concept of instincts as innate, automatic, and unmodifiable patterns of behavior. For example, she claimed to be applying to "human behavior" the "new ethological view of instinct" and cited Robert Hinde as its most recent and representative advocate. According to Ainsworth, this view "does not exclude the notion of modifiability." But when she defined the concept, there was not much difference between her definition and the old concept defended by Lorenz. An instinctive behavior, she noted, was "a behavior pattern, characteristic of the human species, which emerges without a specific individual learning experience and which can most readily be evoked by an identifiable stimulus configuration ... although it may fail sometimes to be elicited by this stimulus and may sometimes be given in its absence."[55]

Ainsworth thus moved from a classification of attachment behavior based on some limited observations and measurements of children's reactions in the strange situation to the claim that the mother is the cause of that behavior, and then to the claim that this behavior has a biological basis. The previous section showed that the first step, from observing differences in a child's behavior to attributing them to maternal care, was not well supported by the data. The second step, from alleging that maternal care causes a child's behavior to asserting its biological basis, cannot be supported by the kind of data she had either. At best, Ainsworth's work could serve to separate behaviors into different categories. It cannot help evaluate the biological basis of that behavior. As such, her observational and experimental work could neither confirm nor disprove Bowlby's ethological theory of attachment.

Yet, by appealing to the "striking similarities" between the results obtained in Uganda and those from Baltimore, Ainsworth suggested that genetic factors influenced the development of attachment. Whether in a poverty-stricken family in an African village, in a middle-class suburban American home, or in the very strange situation of an infant left alone with a stranger in a psychological laboratory, Ainsworth discerned the same "patterns of attachment behavior." The alleged universality of those patterns could then be used to support the existence of a biological basis. Patterns of behavior that are local are more likely to be the result of cultural practices. Patterns of behavior that are present across cultures are less likely to result from training or social practices. The universality of a behavior is not sufficient to prove its biological basis, but Ainsworth used her work to shore up its probability.

Conclusion

In this chapter I've examined Mary Ainsworth's contributions to Bowlby's ethological theory of attachment behavior, arguing that those contributions need to be placed within the context of their complicated personal and professional relationship. At the end of their careers, both presented their work as a joint venture. My analysis of their correspondence complicates this simple picture. Their "happy partnership" was the result of a tense negotiation in which Ainsworth struggled to maintain her independence and prove the value of her work. She finally carved a niche for herself by convincing Bowlby that presenting herself as his equal and as an independent researcher would be mutually beneficial. She was probably right on this point. For Bowlby, her work provided much-needed empirical backing for his theoretical views. And Ainsworth's empirical work obtained greater visibility through its connection with Bowlby's ambitious new paradigm for child development. But the confluence of their views was the result not of independent research, but of self-interested negotiation.

Ainsworth's support of Bowlby came at a crucial moment in his career, given the intense challenge to the concepts and empirical work he had used to construct his grand synthesis of ethology and psychoanalysis. At least in the United States, Ainsworth's observations and experimental data would become crucial to the increasing acceptance of attachment theory. Appealing to her observations of mother-infant relationships in different cultures, Ainsworth claimed that the patterns of attachment she observed were universal. Believing that this provided conclusive evidence of a biological basis for the mother-child attachment, Ainsworth followed Bowlby in adopting the ethological conception of instinctive behavior. By presenting children's emotional needs as universal, I have argued that she helped shore up the claim that such needs had a biological basis.

But in traveling from London to Baltimore via Uganda, attachment research underwent several important and troublesome displacements. As we have seen, Ainsworth often moved from behavior to feelings, from children to mothers, and from relation to causation without sufficient evidence. In addition, she never clarified what attachment really is. Further, as I pointed out, the type of research she carried out with infants cannot prove her and Bowlby's claims about the evolutionary basis of attachment. In the end, as I have argued, Ainsworth's observational and experimental studies cannot carry the weight of proving the theoretical claims of attachment theory.

CHAPTER EIGHT

Reinforcing Each Other and a Normative View of Nature

FATHER: Scientists are always assuming or hoping that things are simple, and then discovering that they are not.

DAUGHTER: Yes, Daddy.

DAUGHTER: Daddy, is that an instinct?

F: Is what an instinct?

D: Assuming that things are simple.

F: No. Of course not. Scientists have to be taught to do that.

D: But I thought no organism could be taught to be wrong *every* time.

F: Young lady, you are being disrespectful and wrong. In the first place, scientists are not wrong every time they assume that things are simple. Quite often they are right or partly right, and STILL MORE OFTEN, THEY THINK THEY ARE RIGHT AND TELL EACH OTHER SO. AND THAT IS ENOUGH REINFORCEMENT.

—Gregory Bateson, "Metalogue: What Is an Instinct?" (1969)

Introduction

In 1969 Bowlby published *Attachment*, the foundational text for attachment theory or the ethological theory of attachment behavior. The book's central thesis is the same one Bowlby had presented in his 1958 paper "The Nature of the Child's Tie to His Mother": that children have an instinctual need for maternal care and love. However, Bowlby expanded the theoretical discussion, updated several points about biology, and incorporated Ainsworth's empirical work as corroborating evidence.[1] Following historian of science Thomas Kuhn's view that scientific fields are configured around a central paradigm, Bowlby presented the ethological theory of attachment behavior as a new paradigm in child development. Although scientists had criticized many aspects of attachment theory, Bowlby and Ainsworth continued to appeal to ethology to bolster their claims. Aware of his debt to ethology, Bowlby wrote in a copy of his book he gave to Lorenz: "For Konrad, who set this ball rolling. From John."[2]

This chapter shows how Bowlby and Lorenz supported each other, creating an impression of interdisciplinary strength and providing legitimacy to their views on children's needs for maternal love and care. By the mid-1960s, both Lorenz and Bowlby faced important challenges to the empirical basis of their theories. Nevertheless, they continued their cross-disciplinary alliance and elaborated on the serious social implications of their positions.

In addition, I focus on one important development: Bowlby's introduction of the concept of the environment of evolutionary adaptedness (EEA). Expanding Lorenz's view of the existence of a preadaptation between specific environments and behaviors, Bowlby postulated the EEA as the environment in which evolution had already designed stable behavioral systems. Because of that, changes in the environment could lead to pathologies, with important biological and social consequences. This concept thus becomes a diagnostic tool to assess what behaviors human societies can or cannot modify. Specifically, Bowlby, like Lorenz, used this conception of evolution to argue against social changes that could disrupt the mother-infant dyad.

Lorenz Appeals to Psychoanalysis

In the late 1950s and the 1960s—although animal researchers had thoroughly criticized his program, as we saw in chapter 4—Lorenz continued to argue for the unification of animal and human psychology. In doing so, he highlighted the relevance of ethology by writing about the instinctual basis of human behavior and its social policy implications.

Lorenz now focused on an issue of great social concern during the Cold War era: the causes of aggression. He never carried out the experiments on primates that he mentioned at the 1953 WHO conference, but using his earlier studies and his knowledge of a variety of animals, mostly birds, Lorenz asked whether humans are aggressive by nature in *On Ag-*

gression, which appeared in English in 1966.[3] In this provocative book, Lorenz defended the idea that intraspecies aggression is a valuable part of human nature: "The aggression of so many animals toward members of their own species is in no way detrimental to the species but, on the contrary, is essential for its preservation." Lorenz pointed to the intrinsic bond between aggression and the maternal instinct, arguing that aggression is necessary for sexual selection and parental care. But to prevent the unbridled natural drive to aggression from becoming detrimental for the group, different species had also developed ways of inhibiting or redirecting it. For the human race, the Cold War conflict presented special challenges. In a society that possessed nuclear weapons and that he thought suffered from disrupted attachments, Lorenz warned about the great danger that humans would override their innate inhibitions against killing members of their own species.[4]

The book focused on aggression, but Lorenz emphasized that his views applied to other social behaviors as well: "Aggressive behavior and killing inhibitions represent only one special case among many in which phylogenetically adapted behavior mechanisms are thrown out of balance by the rapid change wrought in human ecology and sociology by cultural development."[5] Lorenz's larger message here recalled his earlier warnings about the peril of ignoring human instincts for the preservation of the species.

Using the same framework to explain social behavior that he had developed in the 1930s, Lorenz underlined the spontaneity of instincts and their power. In his previous writings, he had highlighted how instincts cannot be modified by the environment. Now he emphasized that the right environmental inputs are essential for instincts to operate. He discussed how instincts work "under normal environmental conditions" and underscored that "innate behavior mechanisms can be thrown completely out of balance by small, apparently insignificant changes of environmental conditions." Among the environmental conditions he pointed to were disruptions in the patterns of attachment between mothers and children.[6]

Lorenz presented his views on aggression as the absolute truth, and the book became a best seller, but many scientists criticized his generalizations about violence and aggression in human societies and his selective extrapolations from various animals to complex human issues. In the book's preface, Lorenz said he was going "to depict the current situation of mankind objectively, somewhat as a biologist from Mars might see it." His views, he added, resulted from impartial inquiry into the natural causes of human behavior.[7] But other animal researchers such as T. C. Schneirla criticized *On Aggression* in print, and many attacked it even more vigorously in private. The anthropologist Ashley Montagu published a collection of fourteen critical responses from important figures in various fields, including anthropology and psychology.[8]

In an interesting twist, however, Lorenz now appealed to psychoanalytic work to support his views about human biological instincts. In the middle to late 1960s, Lorenz started to cite the work of Bowlby, Spitz, and psychoanalysts in general as confirmation of his views. As I noted in chapter 2, during several trips to the United States, he had established good rapport with psychoanalysts and psychiatrists of psychoanalytic orientation, and he realized he would benefit from being connected to this field. In his book on aggression, he emphasized the "unexpected correspondences between the findings of psychoanalysis and behavioral physiology, which seemed all the more significant because of the differences in approach, method, and above all inductive basis between the two disciplines." He also supported the extrapolation of his work to understand the mother-infant relationship in humans by noting the "legitimate use of [my] work done by John Bowlby, René Spitz, etc."[9]

Spitz, who reported having "known and admired" Lorenz since 1935 and having become friends with him "around 1952," presented himself as following "respectfully in the trail" that Lorenz was "blazing."[10] As we saw in chapter 3, Spitz had received devastating criticism from psychologist Samuel Pinneau. Despite those objections, his 1965 book *The First Year of Life* postulated that "*in infancy damaging psychological influences are the consequence of unsatisfactory relations between mother and child.*" Based on the premise that the mother's personality was dominant in the dyad, he claimed that when things do not go right, "the mother's personality acts as the disease-provoking agent, as a psychological toxin." A table identified the psychotoxic diseases of infancy that arose from maternal attitudes. Following Margaret Ribble, Spitz argued that "overt primal rejection" could lead to coma in newborns, that primary anxious overpermissiveness led to three-month colic, that maternal anxiety caused infantile eczema, and so on. Partial emotional deprivation led to anaclitic depression, while complete emotional deprivation led to marasmus.[11]

Although Spitz was critical of Bowlby's reduction of psychoanalytic drives to biological instincts, as we saw in chapter 4, he too appealed to biology to validate his views as Bowlby and Ainsworth did. Like Bowlby, but without referring to him, Spitz adopted the concept of the organizer from embryology, where it refers to the pacemaker for particular developmental processes, and argued that "*concomitant critical nodal points* were operating also in the *psychic* development of the infant." Animal experiments confirmed the existence of "critical phases" in development, he added. Appealing to the authority of biological research, he reasserted his earlier views about the emotional needs of infants, as supported by his "longitudinal studies on several hundred infants."[12]

Lorenz turned to psychoanalysts like Spitz and Bowlby to support his own views on human behavior. In the preface to a collection of his papers from the 1930s reissued in English in 1970, Lorenz noted that the existence of analogies between the social behavior of birds and humans "was maintained in 1910 and everything which ethology has brought to light since, and particularly that which resulted from *the progressive synthesis of ethology and psychoanalysis*," confirms those views.[13]

Lorenz always talked as if his ideas were widely accepted among biologists, and he believed that only those wearing ideological blinders questioned him. In the same preface, he asserted the validity of his early ideas and his newer ones about the instinctive nature of human aggression. He wrote that biologists "treated as a matter of course" the instinctual basis of human behavior. But in drawing political inferences from his theories, he claimed to have incurred "the fanatical hostility of all those doctrinaires whose ideology has tabooed the recognition of this fact." He thus dismissed his opponents' criticisms as ideological while presenting his own views as the "public property of biological science since *The Origin of Species* was written."[14] His appeal to psychoanalysis, however, reflected how little support he had from animal researchers.

Despite Lorenz's claim to possess the unbiased truth, and in contrast to his increasing popular fame during the 1960s, by the mid-1960s few scientists shared his vision of animal studies or his conclusions about animal or human ethology, as we saw in chapter 4. Lorenz himself was painfully aware of this. In 1964 he wrote to one of his few supporters, the University of Chicago psychologist Eckhard Hess:

> The influence of behavioristic thinking in Ethology is unfortunately not decreasing, as I had hoped. . . . You have always believed that that infectious disease was dangerous and have emitted loud warning signals. In reality, I write today only to tell you that among all the ethologists, including unfortunately my dear friend Niko Tinbergen, you are the only one who I consider a real and outspoken comrade.[15]

In the introduction to the edited collection of his essays mentioned previously, it is revealing that Lorenz cited support only from Bowlby, Spitz, and Philip Wylie. In fact, to bolster his position Lorenz mainly referred to Wylie. Notorious for lambasting American mothers in his 1942 best seller *Generation of Vipers* that we examined in chapter 1, Wylie was a very successful fiction writer and essayist, but he had no standing in science. A Cold War warrior with a sharp tongue and strong political beliefs, Wylie found Lorenz an eager ally in the fight against what Wylie called the LIE, or Liberal Intellectual Establishment. Lorenz dismissed critics like the anthropologist Ashley Montagu, who challenged his views about the biological basis of aggression, as "the silliest asses." Wylie assured Lorenz that he shared the experience of having been put down by "the similar gaggle of micro-mongolio-hydrocephallics [sic]."[16]

Pumped up by this correspondence and a visit from Lorenz, in 1968 Wylie published *The Magic Animal*, which he presented as a "companion-piece" to his earlier *Generation of Vipers*. For Wylie the rise of behaviorism had led to a period of folly, fathered by John Dewey and ending in Marxism. Fortunately, he noted, Lorenz and his studies of imprinting offered a healthy alternative and proved that "animal personality, or something like it, was programmed at the start of existence, for ducks, at least; and for many other creatures." Thus, Wylie concluded that ethology, "by demonstrating, for all time, the reality of instinct, returned behavioristic concepts to their proper and special place."[17]

Encouraged by his interactions with Wylie and other enthusiastic supporters including some psychoanalysts, Lorenz became more daring in his pronouncements about the instinctual nature of human social behavior and the social costs of deviating from nature's path. Despite all the criticism his research program had received, Lorenz still insisted on the "genetic programming" of "highly complex behavior patterns" such as animal and human social relations.[18]

Updating his vocabulary, now Lorenz appealed also to the authority of areas of biology that were making important strides at the time. He couched his position in the language of molecular biology and asserted that "not only the apparatuses for sensory perception and for logical thinking" but also "the complicated feelings that determine our interhuman behavior" were "based on genetic programming." The "fixed motor patterns" that constitute the building blocks of behavior "obey Mendel's law, just like the colors of Mendel's peas." These developments in the study of behavior were inextricably linked to molecular genetics, to the point that "many of the discoveries of ethology would be hardly credible without the vast advances that have been made in molecular biology, with all that they reveal about the transmission of data in the germ plasm."[19]

This was partially true, and like many partial truths, this one threw more shade than light. While it was true that molecular biology had made striking advances, it was not true that those advances had illuminated the genetic basis of human behavior. In fact, none of the advances in molecular biology by this point were even related to the issue. But by packaging his views in that language, Lorenz presented his ideas as validated by new and exciting advances in biology. In addition, he appealed to the new conception of genes as carriers of information to support his notion that instincts were not modifiable by experience and the environment. For Lorenz, genes already carried all the information necessary for instinctual behavior. As Evelyn Fox Keller has noted, the metaphor of DNA as carrier of genetic information, as a program, or as instructions for the organism became very powerful: "Although it permitted no quantitative measure, it authorized the expectation—anticipated in the notion of gene action—that biological information does not increase in the course of development: it is already fully contained in the genome."[20]

Here Lorenz adopted the same strategy that Bowlby used in appealing to ethology: cashing in on the authority of a more established or successful scientific discipline. On the one hand, Lorenz often presented himself as the discoverer of new ideas. With the rise of genetics and new developments of molecular biology, he also claimed that his views were justified by the new discoveries in molecular biology. Those discoveries, however, remained in the realm of biochemistry and cellular biology. The growth of molecular biology was spectacular during this period, but its advances were a long way from illuminating the genetic basis of human conduct.

Nevertheless, Lorenz managed to convince many in the media that the scientific community accepted his views and that his few opponents had a political ax to grind. In a comprehensive profile of Lorenz published in the *New Yorker*, journalist Joseph Alsop claimed that Lorenz had not received due recognition for political reasons, especially in the United States. As he saw it: "In reality, Dr. Lorenz and all the other ethologists I have encountered are both liberals and humanitarians. Yet their discoveries have not only upset applecarts in other disciplines; they have also raised doubts about the optimistic humanitarian liberalism that now prevails in many university circles here in America. Thus, much of the criticism of Dr. Lorenz has heavily moralistic—hence nonscientific—

overtones." Alsop, however, was uninformed, since he claimed that "Dr. Lorenz's seriously hostile critics include almost no ethologists, who are, after all, the people who do the work and know the data of their own discipline." Regarding *On Aggression*, Alsop further claimed that ethologists "have strongly affirmed the over-all thrust of Dr. Lorenz's argument." As we saw earlier, this was not true.[21]

In his conversations with the media, Lorenz often focused on the implications of his views for social and family arrangements. He claimed that one had to support the instinctual basis of family life for the sake of society. As he put it in an interview with the *New York Times* in 1970, "The survival of society at all—of human society—is in doubt, particularly if the family structure is not kept up. I believe that the innate program of the human individual is such that he cannot deploy all his possibilities and evolve all his inherent faculties unless it's done within the frame of the normal family." The title of the article summarized his position: "The Family, to Lorenz, Is All."[22]

In another interview he made it clear that when he spoke about the family, he meant the patriarchal family: "I would venture to say that in man there is a direct correlation between the hate among children and the lack of a dominant father." Hostility in the United States "between brother and sister" was due to the lack of a strong father in the American family, he claimed. Again, animals provided the *proof* for such an assertion: "In wolves, for example, when the alpha animal disappears, hostility develops among the inferior wolves. Battles for superiority immediately break out among the young."[23]

Yet Lorenz did not provide any data to back up his assertion that there was either hostility between brothers and sisters in the United States or a correlation between hatred among children and lack of a dominant father. He also provided no justification for the analogy between men and wolves. There are plenty of animal societies in which the social order does not break down in the absence of an alpha male. Just as he did in *On Aggression*, Lorenz chose examples from animal societies that corroborated his social views. In a revealing examination of his claims in this book, Richard Burkhardt has shown that Lorenz was highly selective in using certain examples and sometimes distorted other researchers' positions.[24] In none of his works did he ever explain why humans should sometimes be compared to pigeons, or at other times to wolves, or to any other creatures in the animal kingdom.

It is hard to understand why Lorenz did not realize that his extrapolations needed justification. His friend and fellow ethologist Eckhard Hess once recounted Lorenz's reaction after one of his talks on imprinting in geese:

> After a talk in which he discussed early effects of experience and later develop-ment of behavior in geese, he discussed imprinting effects or lack of imprint-ing effects. A member of the audience asked: "Dr. Lorenz, what can we learn about human behavior from what you have just told us about the behavior of geese?" And Konrad Lorenz answered immediately, without hesitation: "Abso-lutely nothing."[25]

However, by the end of his life, he was arguing not only that his research allowed for objective pronouncements about humans, but also that we should study animals because they provide clues for understanding humans. For this, he argued, geese were the perfect model, and he did not shy away from drawing the implications from his own work. "Imprinting is not limited to birds," he asserted in his interview with Alsop for the *New Yorker*. "Especially in the higher animals, it is obvious that normal rear-ing plays a decisive role in producing a normal individual, and no doubt imprinting is part of this process." Again, he reported that scientists in other communities appreciated his work: "I must say, the psychologists, and even the psychiatrists, have finally begun to be much interested by this imprinting phenomenon."[26]

In one of those humorous coincidences, the same *New Yorker* issue contained a long review of Philip Roth's *Portnoy's Complaint*. A rising Jewish writer, Roth wrote about the peculiar sexual fantasies of a young Jewish man. Portnoy's sexual deviations and his failure to establish a ma-ture relationship are the direct result of his mother's overbearing charac-ter. At the start of the novel, Roth included a mock dictionary definition of Portnoy's affliction, including a citation to the relevant literature: "Port-noy's Complaint: . . . (Spielvogel, O., 'The Puzzled Penis,' *Internationale Zeitschrift für Psychoanalyse*, vol. XXIV, p. 909) It is believed by Spiel-vogel that many of the symptoms can be traced to the bonds obtaining in the mother-child relationship."[27]

The fictional researcher O. Spielvogel must have been a follower of Lo-renz, since he was an expert on birds and the authority behind the move from birds to humans, from ethology to psychoanalysis. Did Roth also in-tend the play on the word *Spielvogel*? In German, *spiel* means "game" and *vogel* means "bird."

Birds had been at the center of the work on imprinting that Lorenz and Bowlby had used to posit a tight link between their views about animal behavior and research on the human infant-mother dyad. By providing support for each other, they reinforced a position about infants' needs at a time when that position received extensive criticism from researchers in various areas: psychoanalysis, psychology, and animal research.

In 1973 Lorenz, Niko Tinbergen, and Karl von Frisch were jointly awarded the Nobel Prize in Physiology or Medicine. They received this most prestigious of scientific awards "for their discoveries concerning organization and elicitation of individual and social behaviour patterns." According to the communication from the Nobel Institute, their first discoveries "were made on insects, fishes and birds, but the basal principles have proved to be applicable also on mammals, including man." In discussing their "discoveries concerning organization, maturation and elicitation of genetically programmed behaviour," the committee included "the behaviour of a mother to her newborn child."[28] What an awfully simple way of summarizing a complex and contested history!

With the recognition provided by the Nobel Prize, the few qualms Lorenz may have still felt about the validity of extrapolating from his research on animal behavior to complex social issues evaporated. In 1974 Lorenz published in English *Civilized Man's Eight Deadly Sins*. Reminiscent of eugenicists' texts from the turn of the twentieth century that warned of the need for selective breeding to prevent the decline of Western civilization, Lorenz used alarmist rhetoric to proclaim the perils of, among other things, "genetic decay." Whereas the eugenicists were concerned about people with low intellectual capacity, Lorenz worried about "emotionally defective people." Among the major suspected causes of degeneration was, as it had been at the turn of the century, the mother of the race.[29]

Other researchers also emphasized the social consequences of disrupting the mother-infant dyad in an increasingly alarmist tone. In his later writings, Spitz, for example, became more concerned about the social impact of deterioration in the mother-infant bond: "From the societal aspect, disturbed object relations in the first year of life, be they deviant, improper, or insufficient, have consequences which imperil the very foundation of society. Without a template, the victims of disturbed object relations subsequently will themselves lack the capacity to relate." Again with inflated rhetoric, he asserted: "They are emotional cripples; more than a century ago jurisprudence coined the now obsolescent term 'moral insan-

ity' for these individuals." Spitz further predicted that emotionally deprived babies would become the criminals of tomorrow: "Deprived of the
affective nourishment to which they were entitled, their only resource is
violence. The only path which remains open to them is the destruction of
a social order of which they are the victims. Infants without love, they will
end as adults full of hate."[30] In sum, a breakdown in the natural order at
the family level causes social decay. When Lorenz visited Spitz's group in
Colorado in 1970, Spitz declared that only people like Lorenz could save
the world from a foretold fatal decline. He did not mince words: "Armageddon is upon us."[31]

Referring to Spitz's work, Lorenz also argued that "lack of personal
contact with the mother during earliest childhood produces—if not still
worse effects—the inability to form social ties, with symptoms extremely
similar to those of innate emotional deficiency." He warned as well that
the limited role of natural selection in modern society allowed the reproduction of "hereditary instinct defects." He claimed it was "difficult to
argue with those who believe that we are living in the days of antiChrist
[sic]," perhaps recalling his friend Spitz's claim that Armageddon was
upon us. Lorenz's tone was no milder: "There is no doubt that through
the decay of genetically anchored social behavior we are threatened by
the apocalypse in a particularly horrible form."[32] And a few years later he
traced this apocalyptic threat to the mother: "The capability of creating
personal ties is atrophying," and the cause lies in the scarcity of mother-
child contacts in early infancy because "mothers don't have enough time
to play" with their children "or don't think this is important."[33]

Clearly, Spitz and Bowlby influenced Lorenz as much as he influenced
them. Lorenz appealed to them to validate his views, and they appealed to
his views to validate their theories on the mother-child dyad. In his 1969
book *Attachment*, Bowlby continued to rely on Lorenz's ideas.

Bowlby Appeals to Ethology

In 1969 Bowlby published *Attachment*, the first volume of a trilogy about
attachment and loss. Slightly revised in 1982, *Attachment* became the
foundational text for the ethological theory of attachment behavior and
has provided the theoretical framework for attachment research to this
day. The book is an expanded and updated version of the ideas Bowlby
first presented in his classic 1958 paper "The Nature of the Child's Tie to

His Mother." The basic thesis remains the notion that the child has an innate need for mother love in order to become an emotionally mature adult. In addition, Bowlby introduced some modifications in his evolutionary framework that have some new and profound consequences.

Bowlby updated some of the notions he had imported from biology and also couched his views in the language of cybernetics that was prevalent at the time. Now he argued that at between nine and eighteen months, the infant's patterns of behavior that contribute to attachment are organized as a cybernetic system of equilibrium, a regulatory system that tends to maintain the infant close to the mother. He labeled this new packaging of his views the control theory of attachment behavior. To support his views, he still mainly appealed to ethology and animal research.

In his *Attachment* book, Bolwby did not provide additional research or observations on children to support his theory. Besides his references to Ainsworth's work in Uganda and Baltimore, his main propositions rested on the observational work James Robertson had carried out while employed at the Tavistock Clinic. At the beginning of the book, Bowlby acknowledged how much he owed to Robertson: "Since the main data on which I have drawn are those of James Robertson, my debt to him is immense."[34] Yet the relationship with Robertson had become strained. For a number of years, Robertson had been uncomfortable with Bowlby's extrapolation of the data he had obtained in hospitals to everyday situations. Bowlby was aware of this, since it was a frequent point of discussion in their meetings about the book they had been writing together for some time. However, he did not discuss it in print. Robertson did not mention it in his publications for many years either. But in a 1971 paper coauthored with his wife and collaborator Joyce Robertson, he explicitly stated his differences with Bowlby:

> In developing his grief and mourning theory, Bowlby, without adducing non-institutional data, has generalized Robertson's concept of protest, despair, and denial beyond the context from which it was derived. He asserts that these are the usual responses of young children to separation from the mother regardless of circumstance.... Our findings do not support Bowlby's generalizations about the responses of young children to loss of the mother per se; nor do they support his theory on grief and mourning in infancy and early childhood.[35]

In this way, as Anna Freud had done earlier, Robertson not only stated his disagreement about the effects of separation in children; he also made it clear that Bowlby had not carried out observational work and therefore did not have his own body of data to back up his theories. In addition, Robertson reminded the readers that he had been the originator of the concepts of protest, despair, and denial (which Bowlby changed to protest, despair, and detachment). It is beyond the scope of this book to examine the complex relationship between Bowlby and Robertson. For our purposes here, the main point is to note that in developing his theoretical framework, Bowlby had relied on limited empirical support from research on children.

To support his views, Bowlby appealed to research on monkeys, although in a selective manner. Bowlby referred mainly to Harlow's early results, to the publications where Harlow discussed the significance of the mother-infant attachment. As late as 1980, Bowlby continued to link the evolution of his thought to Harlow's findings: "Furthermore, confidence that we are on the right track has been enormously enhanced by experimental studies of rhesus monkeys. Harlow (Harlow et al. 1966), for example, found that females which had been deprived of mothering during their own infancy grow up to be mothers which not only neglect their infants but violently reject them."[36] He was convinced that his ideas had become less controversial "thanks in large part to the related studies of rhesus monkeys undertaken by Harry Harlow in the U.S. and Robert Hinde over here."[37]

Bowlby tried to enlist Hinde in his cause, but Hinde's work does not support many of Bowlby's views. Hinde studied mother-child relationships in a group, not as isolated mother-infant pairs. In the rhesus groups, he showed that the most disruptive effects occur when the mother is taken out of the group. When she returns she has to reestablish her position among the members. This disruption affects the relationship between mother and infant more than the separation of the infant from the mother. Thus Hinde's research really showed the significance of social structure in mediating the mother-child relationship.[38]

Still, Hinde was Bowlby's friend, a paid consultant to Bowlby's unit, and Bowlby helped him obtain support for his primate studies. In his writings, Hinde always placed his objections to the ethological theory of attachment alongside much praise for Bowlby's interdisciplinary efforts to explain social behavior. Sometimes Hinde even presented his studies of separation in monkeys as evidence that "adds force" to Bowlby's view that separation could have a detrimental effect on humans.[39] But his research did not support Bowlby's and Ainsworth's views about the existence of a ready-made instinctual system of attachment.

Although with some modifications, Bowlby's theoretical framework still rested upon the main concepts he had borrowed from Lorenz: imprinting and instinct. In *Attachment*, he resorted to imprinting to explain the tie of children to their mothers. Although in an earlier scientific meeting he had recognized that animal imprinting could not be extrapolated to humans, he now equated attachment with imprinting. In his words: "So far as is at present known, the way in which attachment behavior develops in the human infant and becomes focused on a discriminated figure is sufficiently like the way in which it develops in other mammals, and in birds, for it to be included, legitimately, under the heading of imprinting."[40]

Bowlby also claimed that "the patterns of human behavior" resulting in "attachment of young to parents" are "an instance of instinctive behavior."[41] Influenced by Hinde, Bowlby had earlier abandoned Lorenz's hydrodynamic model of energy to explain the functioning of instincts. Commenting on an earlier manuscript of *Attachment*, Hinde complained to Bowlby that he often sounded too much like the early Lorenz and encouraged him to rethink his concept of instinct. For example, in a 1965 letter, Hinde wrote to Bowlby: "In one or two places you talk about the learned components as though they were added on to 'instinctive behaviour.' This is Lorenz 1935."[42] Bowlby then specified that he was not using the concept of instinct as behavior not modifiable by experience. As Ainsworth did as well, Bowlby recognized that experience and environmental factors play a role in the development of innate or instinctive behavior. However, Bowlby also introduced another concept that constrained this apparently interactionist vision of environmental and genetic factors in human behavior.

In his 1969 book Bowlby introduced the notion of the "environment of evolutionary adaptedness" (EEA). He defined the EEA as the normal environment, the one humans evolved in and are adapted to. Although he continued to speak about instinctive behavior, he said he would rather refer to attachment behavior as "environmentally stable" responses that had been organized by natural selection and are phylogenetically adaptive. He proposed that babies are born with a repertoire of responses that evolved through natural selection in the human "environment of evolutionary adaptedness."[43] Thus, although Bowlby now defined instincts as environmentally stable responses, his view did not imply that behavioral responses resulted from the interaction of environmental and genetic influences. Indeed, the environment had played a role only in the past, during the period when evolution by natural selection constructed the best design for attachment.

Bowlby's colleague Mary Ainsworth also adopted this view of attachment. Her 1969 paper "Object Relations, Dependency, and Attachment: A Theoretical Review of the Infant-Mother Relationship" presented "a theory based on ethological principles." Here she noted that "the evolutionary approach implies that when a species has become adapted—genetically adapted—to an environment there is an interlocking between the developmental potentialities incorporated in the genetic code and the potential impingements of that environment."[44]

Their idea that there is an "interlocking" of the innate aspects and the environment, and that the environmental inputs allow the genetic components to develop was imported from Lorenz's vision of organisms. As we saw earlier, in response to Daniel Lehrman's objections that he did not recognize the role of environment in ontogenetic development, Lorenz presented a position in which the information necessary for a trait to develop is contained in the genome. The right environmental inputs then allow that information to unfurl in the natural, and therefore normal, way. For Lorenz it was important to separate that natural design, the normal state, from the pathological.

The concept of "normal" is one of the most difficult things to define in the whole of biology, Lorenz claimed, and is as indispensable as its counterpart, the concept of "pathological." According to him, the test for finding out whether a behavior is normal is to ask: "Is this how the constructor meant it?" That is, the normal is what has been selected for.[45] Because natural selection has already designed a set of behavioral systems that work, interference with that design, including environmental changes, can only lead to malfunctions.

Bowlby adopted a similar normative model of evolution and a similar vision of an organism's behavior. In a way, his notion of the EEA served as a thought experiment to figure out what natural selection (Lorenz's constructor) had designed. In the case of attachment, natural selection had "interlocked" a mother with her infant. For Bowlby and Ainsworth, in the EEA the attachment system was configured to tie the mother to her child. As Ainsworth put it:

> Ethologists hold that those aspects of the genetic code which regulate the development of attachment of infant to mother are adapted to an environment in which it is a well-nigh universal experience that it is the mother (rather than some biologically inappropriate object) who will be present under conditions which facilitate the infant's becoming attached to her.[46]

In this new presentation of the biological basis of attachment, Bowlby and Ainsworth did not merely "update" their vocabulary by packaging their views of attachment in the language fashionable at the time—genetic codes, regulation, and the information metaphors that were prevalent at the heyday of exciting new advances in molecular biology—as we saw that Lorenz also did. In my view, their conception of attachment as a behavioral system already designed in our ancestral past had three important ramifications.

First, this view of attachment as a behavioral system established in the EEA amounted to a reification of attachment. Bowlby posited the existence of an object that had previously not been part of the psychological world: a behavioral system "built in and inherited," also present universally, and thus part of human nature. According to Bowlby, the attachment system comprised infants' and mothers' behavior and feelings, one being the initiator or releaser and the other being activated by the infant's conduct. Although Sigmund Freud, Anna Freud, and other analysts like Margaret Ribble had talked about the infant's attachment to the mother, they referred to a relational property, not to a component of humans' inborn equipment. For them, a mother and her child are active agents in constructing a relationship; in building that relationship, the child learns how to love. For Bowlby, both child and mother come into the world with specific innate needs and behaviors. Although their behaviors also affect the development of the attachment, the mother and the child are merely acting out the predetermined program established by evolution in the EEA.

Second, Bowlby's use of the concept of environment of evolutionary adaptedness to establish a nexus between biological instincts and a specific environment led to naturalizing nurture because the right environmental inputs for the attachment system have already been established. As part of a behavioral system that was an adaptation, the mother became an element of the child's natural, required environment.

Third, this view of attachment as a system already constructed in the EEA gave attachment a prescriptive character when it was combined with Bowlby's view that "environmentally stable behavioral systems are as necessary as morphological structures for the survival of each species."[47] According to this reasoning, natural selection has already designed the appropriate responses for a given environment. Therefore not only genetic changes but also environmental changes will alter the balance established by evolution. Any disruption in those responses or in the environment would thus lead to behavior that might imperil the survival of the species. Because in Bowlby's view the mother is the child's natural "environment," her responses must not deviate from the range of behaviors designed by natural selection and needed for the survival of the child and the species.

For Bowlby, this evolutionary elaboration of attachment had clear social implications worth highlighting in a separate section.

Normative Nature: From the Natural to the Social

As we saw in earlier chapters, Bowlby had long been concerned about the social order. In his early writings he emphasized that lack of maternal care caused psychological problems to infants and had destructive social consequences. "The proper care of children deprived of a normal home life," he stated in his 1953 WHO report, is "essential for the mental and social welfare of a community." Further, "When their care is neglected, as happens in every country of the Western world to-day, they grow up to reproduce themselves. Deprived children, whether in their own homes or out of them, are the source of social infection as real and serious as are carriers of diphtheria and typhoid."[48] Recall also his analogy between the monotropic mother-infant relationship and the British royal order in his 1958 paper in which he presented the core ideas of the ethological theory of attachment behavior.

Bowlby, as I explained earlier, had been criticized for extrapolating the consequences of children's deprivation in hospitals to normal circumstances. In addition, the little research done on working women did not support the view that they harmed their children. But Bowlby stuck "to his guns," as one article reported in 1965. This article presented Bowlby's position as well as Anna Freud's criticism of those who blamed mothers for anything and everything that went wrong with a child. Bowlby dismissed such criticism as coming from "vested interests":

> Whenever I hear the issue of maternal deprivation being discussed, I find two groups with a vested interest in shooting down the theory. The Communists are one, for the obvious reason that they need their women at work and thus their children must be cared for by others. The professional women are the second group. They have, in fact, neglected their families. But it's the last thing they want to admit.[49]

FIGURE 8.1. Cover of John Bowlby's *Attachment*. Courtesy of Basic Books.

Like Lorenz, Bowlby presented his theory as the disinterested position of science; only self-interested parties rejected it, for nonscientific and socially suspect reasons.

For Bowlby, the social message of his theory was transparent. In his successful book *Attachment*, the message appears on the cover: Let's recover the natural mother (see fig. 8.1). In interviews and other writings, he did not shy away from drawing social policy implications from this research: "Day-care centers are a dangerous waste of time and money."[50] Bowlby did not change his mind about this.

Bowlby had defined the environment of evolutionary adaptedness as the normal environment, the one we evolved in and are adapted to. For Bowlby and Ainsworth, the mother constitutes the "normal environment," as she is the "biologically appropriate" individual for responding to her child's instinctual needs. The notion that the attachment system was already designed in the EEA implied a profoundly conservative position regarding child care.

When applied to child care, Bowlby's assertion that environmentally stable behavior systems are necessary for the survival of each species became a prescription against social change in family structure and parental roles. Since the mother-child dyad is one of those systems, its preservation is necessary for the benefit of the species. In Bowlby's view, the environment must remain stable so that "environmentally stable" behaviors can remain adaptive. Nature's order should decree the social order; indeed, it has already done so. Because he equated the natural with the normal and the good, he saw deviations from the natural as sources of psychopathology and sociopathology. As Bowlby put it, "The more the social environment in which a human child is reared deviates from the environment of evolutionary adaptedness (which is probably father, mother, and siblings in a social environment comprising grandparents and a limited number of other known families) the greater will be the risk of his developing maladaptive patterns of social behaviour."[51]

Deviations from that system are not simply variations, but pathologies. Further, changes in the system established in the EEA not only are pathological for the child but put the survival of the species at risk. The idea here is that if natural selection created a design that has worked for thousands of years, interfering with that design can only lead to malfunctioning of the process.

Noting their agreement with Bowlby, in a 1978 book about the strange situation, Ainsworth and other attachment researchers claimed: "To the extent that the environment of rearing approximates the environment to which an infant's behaviors are phylogenetically adapted, his social development will follow a normal course. To the extent that the environment of rearing departs from the environment to which his behaviors are adapted, developmental anomalies may occur."[52]

Ainsworth was also often conflicted about the implications of attachment theory for child care, and she constantly pointed out that Bowlby did not believe children need their biological mothers. Ainsworth struggled with the issue of who should "mother." In some publications she referred to a mother figure: "Bowlby's ethological-evolutionary attachment theory implies that it is an essential part of the ground plan of the human species—as well as that of many other species—for an infant to become attached to a mother figure. This figure need not be the natural mother but can be anyone who plays the role of principal caregiver."[53] However, in other places Ainsworth claimed: "The behavior of the securely attached infant and his responsive mother, in both familiar and unfamiliar surroundings, may be recognized as the expected evolutionary outcome of infant attachment and attachment behavior and of a reciprocal maternal behavior system which are preadapted to each other."[54] Now, if what she called the "dyad" is preadapted, then the maternal "figure" can only be the real mother.

In defending their views, Bowlby and Ainsworth tried to have it both ways, by claiming that evolution had designed a system between mother and infant and by asserting that the person taking care of the child could be someone else. But first, they never did any research with other subjects—aunts, uncles, fathers, or any other people who interact with children and take care of them. Despite the occasional disclaimer about mother figures, they almost always talked about the biological mothers and they never explained why their views about them should also apply to other caretakers. Second, as Bowlby and Ainsworth anchored their views deeper in evolutionary biology, they became more committed to the view that the biological mother is the best caretaker for her child.

In most publications, Bowlby and Ainsworth said that mothering contributed to the survival of the species. Later, when Bowlby became aware that many biologists rejected group selection, he justified the instinctual character of parental care by using a gene-selection model: "When looked at in terms of evolutionary theory, the occurrence of altruistic care of young is readily understood since it serves to promote the survival of offspring . . . and thereby the individual's own genes."[55] But a gene-selection model would explain why one cares for one's own children, but not why one cares for others' offspring. Accepting that the mother is the natural and normal caretaker for her child was simply the price to pay for positing attachment as a biological system.

The presentation of the mother-infant dyad as the design of evolution by natural selection continued to have a strong emotional power for mothers, who would be told their children were designed by nature to need their care and love. Furthermore, I also think it introduced a firmer justification for the authority of nature in social affairs.

In his 1958 paper, Bowlby had basically presented a functionalist justification for the designs of nature and the designs of society. But now he was going further by claiming that natural selection had already established a specific genetic configuration that became adapted to a given environment during the environment of evolutionary adaptedness. The consequences of this conception of evolution are perforce conservative, since any modification of the genetics or of the environment can only disrupt the design already in place.

In fact, Bowlby's position implies what I would call evolutionary determinism. This type of determinism comes in with the notion that human nature is already constructed. The view that natural selection has already constructed certain traits, behaviors, or cognitive structures, molding them into functional designs for a given environment, often leads to the idea that any changes could only be disruptive. Because the environment is now considered part of the evolutionary design, changes in the biology or in the environment threaten to destabilize the functional integration of the adaptation.

Because the environment is constructed as part of the evolutionary design of particular adaptations, Lorenz and Bowlby could emphasize the role of the environment while noting the danger of changing it. However, for them the environment simply decodes what is already inscribed in our genetic heritage.

Conclusion

This chapter has followed the fate of Bowlby's and Lorenz's ideas until the mid-1970s. Lorenz, whose views about human instincts were by now widely criticized, appealed for support to some psychoanalysts like Spitz and Bowlby, and he argued that his views were corroborated by the ongoing synthesis of psychoanalysis and ethology. Bowlby and Ainsworth claimed that ethology supported their views about the infant's instinctual need for maternal care and love while they relied on each other and on advances in genetics that they claimed supported the view that behavior had an instinctual basis. They also sought to delegitimize their critics' positions as ideologies and used Cold War rhetoric to show how their theories were relevant for society's survival.

Following Lorenz, Bowlby and Ainsworth argued that evolution by natural selection had already designed a variety of behavioral systems that require the interlocking of genetic and environmental factors. Because human nature is already constructed, deviations become pathologies. This vision of a ready-made combination of genotype and environment explains why Lorenz worried that insignificant changes in the environment could derail innate behaviors. This concern is congruent with his conception of evolution as a process that has already designed the world with a specific set of matching keys and locks, and with instincts that function properly in a given environment. Bowlby provided further support for this view by introducing the notion of the environment of evolutionary adaptedness, the period during which natural selection had designed the mother-child attachment system.

Bowlby's and Ainsworth's appeal to the biological foundation of the mother-infant dyad required naturalizing environmental inputs that shape an infant's emotional development. By doing this, they constructed a functionalist argument to support the status quo regarding parental roles in society.

Despite the criticisms that the ethological program had received, Bowlby and Ainsworth never ceased claiming that research in biology supported their views. They were aware that the validity of their ideas depended heavily on that support. This is what distinguishes attachment theory from a variety of other psychological and psychoanalytic theories dating to the 1930s that postulated the mother as the cause of her children's ailments. As Ainsworth and Bowlby themselves noted as late as 1991: "The *distinguishing characteristic* of the theory of attachment that we have jointly developed is that it is an ethological approach to personality development."[56]

The appeal to a biological foundation for the mother-child attachment, and to ethology in particular, was crucial for attachment theory. But as this book has also shown, the ethological part of the ethological theory of attachment behavior rested on shaky foundations.

Conclusion

Infants, Instincts, and Mothers

We started with a question: What do children need if they are to develop into emotionally healthy individuals and good citizens? I have explored the history of one response to this question—*mother love*—and have focused on the development of one particularly successful variant of this response: John Bowlby's ethological theory of attachment behavior. According to this theory, in early infancy there are critical or sensitive periods for emotional development that are determinant for adult personality. The infant requires empathic love from the mother, because the mother-infant relationship is a natural dyad designed by evolution in which the instinctual responses of one party activate instinctual responses in the other. Attachment researchers claim that a secure or insecure attachment to the mother affects all other relationships in life. That is, they have turned a view about the child's relationship with the mother into a theory of personality development. According to Bowlby, Lorenz's studies of imprinting, Harlow's and Hinde's primate research, and Ainsworth's observations of children supported his theory of attachment.

Here I have argued that developing and maintaining the theory of attachment behavior was not a simple, straightforward process of putting together data from human infants, ducks, and monkeys that all supported the same views about the child's need for the mother. I situated the rise of attachment theory within more general views about the influence of mother love, the rise of emotions as a key component in the personality of a mature individual and good citizen, and the return of instincts in explanations of human behavior. I have shown that in turning the infant's love for the mother into a biological need, attachment theory built on and contributed to the biologizing of human nature and to a vision of human behavior as preprogrammed by evolution.

In part 1, I showed how Konrad Lorenz and John Bowlby became the main architects of the view that infants have an innate need for a mother's care and love. I followed the move from imprinting in animals to attachment in infants, and with it the construction of maternal love as a biological need. This move built on considerable psychoanalytic research on the mother-infant dyad and was intimately related to post–World War II concerns about how individuals become emotionally sound. I argued that to explain the rise of Bowlby's theory we need to understand the authority conferred on it by Lorenz's work, the widespread preoccupation with the emotions after World War II, and the debate about parental roles in the wake of women's increased access to the workforce. In the context of Cold War debates about gender roles and working mothers, Bowlby's views helped justify the patriarchal family.

In part 2, I showed that Bowlby's project to provide a synthesis of psychoanalysis and ethology received pointed criticism. Researchers from various fields rejected his effort to explain human behavior by biological instincts and his reduction of infants' needs to natural mother love. Comparative psychologists working on animal behavior criticized basic concepts that Bowlby had adopted from Lorenz. Daniel Lehrman challenged the possibility of explaining the behavior of an organism without knowing its ontogenetic history. Other animal researchers also questioned the significance of imprinting and the critical periods hypothesis. In addition, many Freudian psychoanalysts—including Anna Freud, Max Schur, and René Spitz—faulted the reduction of psychological drives to biological instincts. They noted that in constructing the mother-child relationship as instinctual in the ethological sense of the word, Bowlby had eliminated the mental aspects of instincts, which were the focus of psychoanalysis. As Anna Freud pointed out, in explaining the mother-child relationship in terms of biological instincts, Bowlby ignored what matters most to psychoanalysis: the individual's subjective experience. Last, we saw that Harlow's work on rhesus monkeys, contrary to most interpretations, did not support Bowlby's views. Instead, it showed the importance of peers in infant socialization.

All these researchers emphasized the importance of experience, contingent historical events, and individual variability in explaining human emotions and behavior. They all argued against reducing complex human mental and emotional states to preprogrammed biological instincts, as Bowlby and Lorenz had done.

In part 3, I followed the ways Bowlby and Lorenz maintained their views, despite considerable criticism from different quarters. By looking at how Bowlby used Mary Ainsworth's work on infants and how he and Lorenz supported each other, we encountered a new twist in the naturalizing of parental roles. Starting in the mid-1960s, Lorenz, Bowlby, and Ainsworth appealed to the evolutionary significance of the right environmental inputs. Bowlby proposed that the environmental factors that make the mother-infant dyad function well had been selected in our evolutionary past, reinforcing their biological significance. Bowlby added that babies and mothers are born with a repertoire of responses that evolved through natural selection in the human "environment of evolutionary adaptedness," the ancestral environment that human beings evolved in and are adapted to. Because the mother is the child's "environment," in Bowlby's view any deviations from the patterns established by natural selection would increase the risk of developing "maladaptive patterns of social behaviour."[1] In this way the description of children's needs automatically turned into a prescription for maternal care and gendered parental roles.

Although attachment theory emphasized the significance and value of mother love, I have argued that turning love for mother and mother love into biological instincts had negative consequences for mothers. This view gave them a monstrous power to affect a child's life. In fact, as the literature on the power of mother love grew, so did a discourse of blame that held mothers responsible for all of their children's emotional problems. If the mother is the psychic organizer, it follows that children's problems result from the absence of mother love, from the derailment of their mothers' natural instincts.

In addition, by reducing maternal care and love to a matter of instincts, Bowlby conceived of a mother's behavior and feelings as automatic responses to her infant's biological needs. As Bowlby put it, "The normal mother can afford to rely on the prompting of her instincts in the happy knowledge that the tenderness they prompt is what the baby wants."[2] Although this position gives mother love a central role in personality development, I have argued that when considered to be merely a natural feeling, mother love is devalued.

In constructing mother love as a sort of physiological epiphenomenon, many child analysts situated maternal sentiments close to animal-istic mechanical behavior and, subsequently, far from the realm of voli-tion, intelligence, and reason. As such, they left mother love outside the realm of moral praise. Maternal care and love are no longer the result of personal choice, intelligent decisions, or dutiful sacrifice; as with any other biological and automatic emotion, mother love does not deserve moral commendation. Loving her child is just a woman's natural destiny. That is why for these psychoanalysts the good mother is the "unthinking" mother, the woman who merely follows her maternal instinct. To fulfill her role, a mother need not resort to conscious will, personal striving, or thoughtful choices. She can simply function as an animal. Following the dictates of biology allows mothers few options.

Although I have not developed this point in my book, it is worth not-ing that it also allows few options for children. Children have feelings too, and scientific study of them helped recognize their emotional needs. But in turning those emotions into biological needs that can be satisfied only by the right environmental factors—their mothers—attachment theory deprived children of any agency in their life histories and robbed them of any power over their destinies. With their needs determined by evolution, and with the fulfillment of those needs in the hands—or hearts—of their mothers, children seem to be at the mercy of forces beyond their control.

By turning to the emotions as a reaction to the mechanized post–World War II world, these scientists ended up, ironically, with a concep-tion of human beings as machines designed by nature to perform auto-matic functions. According to Ainsworth, Bowlby, and Lorenz, this design must be respected, for disastrous consequences would come from tinker-ing with a design that has proved reliable during our evolutionary history. In the search for greater agency in an increasingly mechanized world, the science of mother love has given us a reductionist and mechanistic con-ception of the human mind and behavior.[3]

*   *   *

Despite the history of criticism of Bowlby's ethological theory that this book has uncovered, attachment theory became one of the most influ-ential psychological theories of the second half of the twentieth century. Although some psychologists have disapproved of certain aspects, in the early twenty-first century attachment research still remains a major area of scientific study, and the field continues to expand. Current work focuses on experimental research on the strange situation, measuring attachment, and evaluating its effects in adulthood.[4] The advance in attachment research is no minor matter: as Diane Eyer notes, "In the twenty years between 1974 and 1994, some 3,535 attachment studies were published and the annual rate of publication grew from 35 in 1975 to 280 in 1993."[5]

Today attachment research continues apace, with several journals and research centers dedicated to it. The theory's success is also visible in its expansion into other areas within psychology, social thought, and general culture. Scientific articles, movies, and newspaper articles refer to the immense literature on attachment, and even to "attachment disorders." Although many psychologists working on attachment have expanded the field to include infants' attachment to other caretakers, the dominant focus is still on mothers.

In the child-rearing literature, the influence of attachment remains alive too. In the 2003 edition of *The Baby Book*, William and Martha Sears write that studies show that *"the most important contributor to a baby's physical, emotional, and intellectual development is the responsiveness of the mother to the cues of her infant."* Although they note that the mother doesn't have to hold the baby all the time, they emphasize that the most important thing is her "responsiveness" and the *"harmony"* between mother and infant. With over half a million copies sold, texts like this one sustain worries about the pathological consequences of deviant mothering. The Searses highlight "the most common infection of employed mothers—*distancing between mother and child*. This occurs as a result of mothers having other thoughts than their children at work."[6] Neither the message nor the rhetorical packaging seems to have changed since the 1950s.

Given the substantive criticism the theory received, one must ask: How could attachment theory survive the criticism and the second wave of feminism in the late 1960s? A full answer is beyond the scope of this book, but I present some tentative ideas. Until the 1970s, the period examined in this book, attachment theory was successful, in my view, for three main reasons: its appeal to the authority of biology; the unified front its leaders presented, in contrast to their critics' dispersion; and the hybrid character of the theory.

The most important reason for the success of attachment theory was its defendants' continuing appeal to the authority of biology. As I explained earlier, the ethological theory of attachment capitalized on the authority of biology at different levels. On a methodological level, Bowlby and

Ainsworth claimed the authority of the methods of ethological research. Ethology managed to present itself as a discipline grounded on observational fieldwork that also provided experimental results for its conclusions. It thus partook of the emphasis on experimental research characteristic of modern science and, specifically, the prominence of experimentation in postwar scientific epistemology. Perhaps because of this, it managed to confer a high status on observational studies. Bowlby and Ainsworth presented themselves as following ethologists by combining the "naturalistic" observation of children and infants with experimentation procedures such as the "strange situation." On a conceptual level, attachment theory obtained the legitimacy of a science with a higher status by adopting concepts like instincts and imprinting, which were part of ethology and for a while had wide currency in postwar scientific and popular discussions of behavior. Finally, on a more personal level, attachment also benefited directly from Lorenz's support. Lorenz agreed that research on animals confirmed the existence of universal needs in infants and supported a biological basis for attachment. Coming from a Nobel Prize winner of international standing, his support for Bowlby's work carried significant weight.

A second factor that played a central role in the success of attachment theory was the ability of its main developers to present a unified front, while their critics failed to do so for a variety of reasons. Bowlby and Ainsworth overcame their differences—intellectual and personal— to join in defending attachment theory. As we saw, by portraying her work as independent confirmation of Bowlby's work, Ainsworth actively developed a strategy that benefited both of them and reinforced the image that their different approaches converged into a unitary position. Bowlby used several strategies to accomplish that goal. He picked and chose from the results of other researchers, such as Harlow and Hinde, those results that agreed with his views. This was recognized even by Ainsworth, who confessed to Bowlby how she admired "your ability to select from many writers points that contribute to your argument, gracefully acknowledging their contribution, without being blocked (as I so often am) by the fact that other things they say in the same paper are contrary to your view."[7] He never acknowledged in print his deep disagreements with Robertson. Robertson eventually published his criticism, but his status as a social worker mainly interested in practical reforms lessened the influence of his views. Finally, Bowlby never substantially engaged his critics. He rarely noted how other researchers challenged or differed from his position, either ignoring them or quickly dismissing their objections.

A contributing factor to the success of this strategy of presenting a united front was the failure of the main critics to do the same thing and engage in a continued effort to challenge the empirical basis and the conceptual shortcomings of attachment theory. Again, this happened for a variety of reasons. Daniel Lehrman, for instance, died in 1974 at the peak of his professional life, when his status and work could have made the strongest impact. Samuel Pinneau and Lawrence Casler, the psychologists who published the most comprehensive critical reviews of the literature on maternal deprivation and Bowlby's work, moved on to other areas of research.

The third reason for the success of attachment theory was that the hybrid nature of the notion of attachment shielded it from a focused critical appraisal. The theory's interdisciplinary character helped it survive because the strict epistemological rules and standards of evidence required within discipline boundaries were difficult to apply in this case. The belief that mother love is essential owes much of its endurance to that hybridity, which made it possible to escape the close scrutiny that scientific statements typically undergo when they rest on research pursued within established disciplines. Even today, many attachment theorists deflect criticism by pointing out that the theory has the backing of biology. Yet biologists have not carried out research that supports Bowlby's and Ainsworth's ideas on the biological basis of behavior.

The history of attachment studies since the 1970s remains to be explored. But I believe any account of their successful expansion and continuing growth must pay attention to two factors: the role of the strange situation in affording a simple experimental procedure that allows fast production of experimental results, and the halo effect of its biological foundation.

The post-1970s expansion of attachment research owes much to the booming success of the strange situation as a laboratory tool. The growth of studies of children in the strange situation is hard to overlook: Robert Karen says it came to be "more widely used than any other in the history of developmental psychology."[8] The use of the strange situation promoted a rapid expansion of experimental research. Attachment researchers produced data, published new results, trained new recruits, and built a large scientific network aided by this new procedure. The strange situation constitutes a central element of the material and social systems of production which, as Robert Kohler has argued in the case of *Drosophila* genetics, "enable experimenters to mobilize material resources, socialize recruits, and persuade other workers to accept their results and adopt their methods of production."[9] Thus the experimental study of attachment has become a thriving industry in its own right. At the same time, the residual connection to biology continues to lend credibility to the field.

The continuing appeal of the biological basis of attachment rests on what I call a halo effect. By now most of the biology Bowlby used to erect his theory has been discredited. Yet attachment practitioners continue to point to the biological foundation of the field. I suggest that by hitching their research program to an established field with higher scientific status, they place attachment in a favorable light that confers higher authority to their views.

Because current researchers still appeal to Bowlby's and Ainsworth's work, a history of their views becomes relevant in assessing the foundations of current attachment theory. Contrary to other histories that present the development of the ethological theory of attachment behavior as a story of increasing accumulation of evidence and consensus building, my investigation has uncovered a more complex story that questions the progressive development of the ethological theory of attachment and raises serious doubts about its validity. I believe my historical analysis amply sustains the conclusion that the scientific evidence in support of attachment theory has been insufficient and is deeply flawed.

Historians of science are often wary of making such judgments, lest our peers think we have abandoned the mantle of neutrality allegedly needed to understand the past. Whereas, based on their historical investigations, political historians can—and do—assert that a president was right or wrong in making a decision, historians of science are often expected to withhold judgment about whether a scientist's claims were or were not supported by the evidence available. I agree that we need to try to understand a theory in its scientific and social context, regardless of whether it is accepted from our present perspective. In doing so, the historian of science works like an anthropologist who tries to understand the practices of a specific culture regardless of how foreign, or even nonsensical, they seem in her eyes. But her findings may let a historian draw some conclusions about the validity of a scientific claim or theory.

In my view, at least four fundamental challenges to Bowlby's position—raised mainly by Anna Freud, Daniel Lehrman, and Harry Harlow—have not been adequately confronted. First, attachment theorists still need to better clarify how the nexus between behavior and emotions is constructed. Until this is done, a central tenet of attachment theory—

the notion that the infant requires empathic love from the mother and not simply care—cannot be fully supported. How do we measure "empathic" love? Though they appealed to the ethological concept of instinct, which focused only on behavior patterns, Bowlby and Ainsworth also argued that doing the right thing without the right feelings is not enough for good mothering. But how do we know an infant can peer into a mother's feelings? And how can the scientist know? As Anna Freud pointed out, one cannot "deduce" the feelings of children and mothers just by observing their behavior.

Second, as Harlow's and Hinde's studies showed, it is important to recognize that infants do not relate only to mothers. If their studies of rhesus monkeys showed anything, it is precisely how complex social animals are: how much play, peers, family structure, and social setting influence their development. If that is true of monkeys, how can the human infant's need for a mother, or even one attachment figure, be established without research on the role of fathers, siblings, peers, and other relationships in the lives of children (not to mention how all those relationships affect one another)? Given the differences he observed in social structure, Hinde increasingly emphasized, contrary to Bowlby's position, that rhesus family groups and human families cannot be compared:

> All societies are complex and diverse. For example, in groups of monkeys the mother-infant relationship is affected by the presence of other individuals, the mother's dominance status, the mother's relationship with the male, and so on. Conversely, the presence of an infant affects the mother's relationship with other adults. . . . That even greater complexities occur within human families and other groups hardly needs saying.[10]

Moreover, Hinde directly criticized several aspects of Bowlby's and Ainsworth's work. He noted their lack of clarity about exactly what behaviors are part of the instinctual system. He also questioned their assumption that infant and maternal behaviors are adapted to "mesh with each other."[11] Yet they never ceased claiming that research in comparative psychology and biology supported their views. In a 1991 piece about the historical development of their theory, Ainsworth still claimed support from Harlow's and Hinde's studies.[12]

Third, the distinctive characteristic of attachment theory—the claim that the mother-infant relationship is a natural dyad designed by evolution in which the instinctual responses of one party activate instinctual responses in the other—remains unsubstantiated. Neither Bowlby nor Ainsworth adequately defended the problematic model of evolution and of the organism that provides the basis of their views about the dyad, a model that appeals to the environment of evolutionary adaptedness, posits the existence of preprogrammed emotional needs, implies the naturalization of nurture, and assumes an ideal balance between our genetic makeup and a particular environment.

Bowlby, similar to Lorenz, adopted what philosopher Elliott Sober has called a "Natural State Model." This is the view that there is one developmental path that counts "as the realization of the organism's natural state, while other developmental results are consequences of unnatural interferences." Underlying this model is an Aristotelian vision of natural states and essences profoundly at odds with Darwinian evolution. Ever since Darwin, evolutionary biology says that no set of traits or behaviors can be taken as the essence of a species. Nor is there a combination of a genotype and an environment that can be taken as the natural state. Lorenz's pre-Darwinian typological and essentialist thinking is revealed in his belief that a "natural environment" determines the one phenotype that is ideal, normal, and natural.[13]

Furthermore, as Daniel Lehrman emphasized, the complex way environmental and genetic factors interact in behavior cannot be elucidated without ontogenetic studies. Lehrman had argued that the environment does not simply decode genetic instructions for behavior patterns that are laid down by our evolutionary history. Nor is the environment's role merely "additive," in the sense of adding to an innate blueprint for behavior; instead, the environment plays an active role in creating the final result.

If attachment theorists want to maintain their appeal to the authority of biology, they need to clarify their claims about exactly how biological research supports the existence of a system of attachment. In 1991, Ainsworth asserted: "The great strength of attachment theory in guiding research is that it focuses on a basic system of behavior—the attachment system—that is biologically rooted and thus species-characteristic."[14] And that is what other supporters of attachment theory also emphasize: "One of the great strengths of attachment theory is that it is based firmly within an evolutionary, ethological and ultimately a biological framework."[15] But "biologically rooted" and "based firmly within" are general statements that do not provide any support for attachment theory specifically. All human activities are "biologically rooted," since we are living beings. If the synthesis with biology is taken as the distinctive characteristic of attachment theory, its theorists need to provide a more specific foundation for their claims.

Fourth, another fundamental unresolved question in this interdisciplinary history concerns the role of animal behavior studies. The question is: How has the use of animal models and research on ducks and monkeys been justified as an appropriate vehicle to understand human emotions? Imagine you open a book about monkeys' development and find that the first article concerns the behavior of lions. Would you be surprised? Surely. Yet it often happens that in a developmental psychology textbook the first article mentioned is Harry Harlow's 1958 paper "The Nature of Love," which reported on experiments carried out on rhesus monkeys. We have become so used to seeing other animals as "stand-ins" for humans that we often do not question or wonder about the epistemological quandaries raised by this scientific practice.

In his last paper, written soon before his death, Lehrman reflected on this key question. "It is sometimes difficult to understand why psychoanalysts who come into intimate contact with the sources of complex human interactions in their practice feel that the experience of people who work with animals is likely to provide a more valid guide to what is happening with their patients than their own experience," wrote Lehrman in "Can Psychiatrists Use Ethology?" presented at a 1970 conference and published a few years later.[16] Here he raised a central issue in the discussions about behavior: Can we really generalize from results obtained with one species to another? More specifically, how much can we learn about human behavior and motivation by studying another species, such as rhesus monkeys or greylag geese? As Lehrman put it,

> What does it mean to interpret the behaviour of one kind of animal by reference to the behaviour of another kind of animal? Can we make interpretations by looking at "equivalent" processes? Can we appropriately and meaningfully look at an event or partial organization in one animal species (e.g., the rhesus monkey) and use that to increase our understanding of an "equivalent" or, for us, a phenomenonally [sic] identical process in another animal such as (to take another example at random) the human?[17]

At the 1970 conference Lehrman asked people to consider the work of I. Charles Kaufman, who was presenting a paper about two species of monkeys, bonnet and pigtail. His work showed that when young bonnet macaques and pigtail macaques are removed from their mothers, the pigtail young do not make contact with another adult, but the bonnet infants do. Thus the effects of separation from the mother are radically different in the two cases. He argued that this was probably due to differences in life patterns and social relationships. Pigtail monkeys are raised only by their mothers, whereas other monkeys in the social group help raise bonnet monkeys. Like Kaufman and Hinde, Lehrman understood that data from animals and humans needed to be interpreted within the context of the different social arrangements a particular species develops in a given ecological context.[18]

This understanding would obviate simple conclusions about what kind of care infants need, because human societies and nonhuman primates organize their lives in various ways, as primatologists like Kaufman pointed out. In reviewing primate parental care, he emphasized the diversity of patterns in the behavior of mothers and fathers. In his own lab, Kaufman reported strong differences between male bonnet macaques and pigtail macaques. Adult male pigtail macaques showed no interest in the young and did not care for them, even when the mothers were removed from the group. The male bonnet macaques played with the older infants, and when the mother was removed they also held, carried, and protected the infants. As Kaufman recognized, it is difficult to make simple generalizations about parental behavior among primates.[19]

The supporters of attachment theory, from Bowlby to contemporary scholars, have not explained how studies on ducks or rhesus monkeys can offer valid interpretations of the role of emotions in human behavior. This doesn't mean they are irrelevant, but certainly they cannot simply be taken at face value. Although some evolutionary psychologists and sociobiologists have also welcomed the ethological theory of attachment behavior, they have not justified using other animals to talk about humans either. Sociobiologist Sarah Hrdy, for example, upholds Bowlby's views as the major breakthrough in evolutionary studies of motherhood.[20] But in *Mother Nature,* her latest major contribution to the topic of mother love and maternal instincts, Hrdy has not addressed some of attachment's shortcomings. Much like Bowlby and Ainsworth, she slides from talking about motives to talking about behaviors and then to talking about biological processes. She also continually cites animal research without explaining why it is relevant to human behavior. For example, Hrdy tell us,

> Sociobiology is not a field known for the encouraging news it offers either sex. Yet its most promising revelation to date has to be that over evolutionary time, lifelong monogamy turns out to be the cure for all sorts of detrimental devices

that one sex uses to exploit the other. As usual, this point is most convincingly demonstrated in organisms that breed much faster than we humans do. Once again, fruit flies are the organism of choice, the current favorites for studying coevolution between the sexes.[21]

It is puzzling to read that research on the behavior of fruit flies will "convincingly demonstrate" anything about human monogamy or the cures for human exploitation.

Even if we could compare some behaviors and emotions between different species, what would those comparisons mean, and which species should we compare ourselves to? Let's focus only on primates. The New World small monkeys of Central and South America include marmosets, tamarins, howler monkeys, night monkeys, spider monkeys, squirrel monkeys, vervets, and titis. Among them, there is great diversity in maternal and paternal behavior. The Old World monkeys in Africa and Asia include langurs, colobus, baboons, and macaques. The langurs rely on alloparenting, where different females take care of infants. In macaques and baboons, mothers take care of their own babies. Males are involved to different degrees, with hamadryas baboons being the most involved and Barbary macaques the least. We are more closely related genetically to the apes of Africa and Asia. These include gibbons, siamangs, orangutans, chimpanzees, and gorillas. The last three are the great apes. In the orangutans, the mother lives with her children and male adults generally live alone. In gorillas, the mother usually does most of the rearing. Chimpanzee mothers take care of their babies for the first six months, then the young began to interact with other kin, which care for them. Needless to say, there is even greater diversity among other animal species.[22]

While the results from animal behavior continue to be fundamental for the ethological theory of attachment behavior, its supporters have not successfully confronted the difficult epistemological problems raised by using results from one species to support views about the social behavior of another.

I have presented four challenges to attachment theorists. Until they address those objections, it seems to me unjustified to assert that children are born with preprogrammed emotional needs established by evolution. For attachment scholars, the mother-child dyad, constructed in the environment of evolutionary adaptedness, has become the optimal form of child rearing. Modern society disrupts this dyad at its peril. Conceived of in this way, our evolutionary history determines or constrains our present choices. Yet this deterministic view of social behavior rests on shaky foundations.

*   *   *

Much is at stake in the naturalizing of emotions, and specifically in the development of scientific views about maternal care and child rearing. Debates about balancing work, child care, and the other life roles and responsibilities of mothers and fathers are central to our culture and to our thinking about the good society. To appeal to instincts as if that were the last word of science, or as if the salience of biology for social life were clear-cut and noncontroversial, closes the door prematurely on important debates about individual and social responsibilities. Raising children is hard work. It requires willingness, intelligence, and rational planning and decision making. No mother can rely on her "instincts," nor should we dismiss the role of fathers or other caretakers.

Naturalizing parental roles also raises the stakes for not following the wisdom of attachment theorists. Women who oppose them have to contend with the authority of science, explaining why their actions would not disrupt the well-being of children, the social order, and the order of nature. If science becomes the arbiter and scientists the judges of the designs of nature, the question of who should raise children is not to be answered by social debate, but by following the dictates of nature. Mothers (or fathers!) will have no real voice in providing the answer.

In this history of mother love and love for mother, we can see that science has become a major player in constructing emotional ideology and in configuring our emotional options. I hope this book has shown how much is at stake when scientists naturalize mother love and turn children's needs into biological instincts.

Acknowledgments

The path to researching and writing this book has been paved with many acts of kindness and many words of encouragement. Although I cannot mention them all, I am pleased to acknowledge some people who helped my journey.

My deepest gratitude goes to my family, for supporting me in many ways and for encouraging me to find my own course. I am most grateful to my mother for the wisdom, energy, hard work, and common sense she has put into loving and caring for all of us. Many, many thanks to my father, my brothers José Luis and Chimo, my sisters Raquel and Silvia, and to Mark for always being there for me; to Erika, Jordi, Ximete, Lara, Adrián, Javier, Marta, and Aitana for many good times; and to Nela, Ximo, Maria, Esteban, and Antonio, who have also helped on many occasions; and thanks to HAT for all the fun.

Early work on this project benefited from many discussions with Peter Galison and Everett Mendelsohn and from their steadfast support. Thanks as well to Richard Lewontin and Lorraine Daston for insightful critical comments. I was fortunate to benefit from the comments, advice, and support of fellow scholars and friends: Richard W. Burkhardt, Luis Campos, Evelyn Fox-Keller, Peter Galison, Ellen Herman, Juan Ilerbaig, Susan Lanzoni, Bernard Lightman, Diane Paul, Victor Pambuccian, Rebecca Jo Plant, Charles Rosenberg, Mark Solovey, Denis Walsh, and Nadine Weidman. For their professionalism and kindness, I thank the team at the University of Chicago Press: Karen Darling, Abby Collier, Erin DeWitt, and Micah Fehrenbacher. I also thank Alice Bennett for her superb editing.

For permission to cite materials or provide images, I thank the copyright owners, libraries, and archives cited in the sources. Special thanks to the archivists at the following archival repositories: the Archives of the

History of American Psychology, the Center for the History of Psychology, the University of Akron; the Western Manuscripts and Archives at the Wellcome Library, London; the Manuscript Division of the Library of Congress, Washington, DC; the Harlow Primate Laboratory, University of Wisconsin–Madison; the Oskar Diethelm Library, DeWitt Wallace Institute for the History of Psychiatry, Weill Cornell Medical College, New York City; the Institute of Animal Behavior, Rutgers University, Newark; and the Konrad Lorenz Institute for Evolution and Cognition Research in Altenberg, Austria. In those places, some people went beyond the call of duty to help me find the materials necessary for my work and to share their knowledge. I am touched by the kindness of Klaus Taschwer, who put all his materials on Lorenz at my disposal, and of Helen LeRoy, who allowed me to use Harlow's papers, shared her invaluable knowledge of Harlow's lab, scanned pictures and sent them to me, and provided kind words of encouragement over the years. I am grateful to Naomi Miller for providing me with a picture of Daniel Lehrman and for encouragement.

I am very grateful to the institutions that make it possible for me to think, study, write, and teach whatever I want, a privilege I do not take for granted. The University of Toronto, Victoria College, and the Institute for the History and Philosophy of Science and Technology (IHPST) provide a most encouraging environment to pursue my work. I am thankful to my IHPST colleagues for their support in the past few years.

For funding for this book project, I thank the Social Science and Humanities Research Council of Canada, the IHPST, Victoria College, and the University of Toronto. Finally, I thank the publishers for permission to use material from the following of my papers: "The Social Nature of the Mother's Tie to the Child: John Bowlby's Theory of Attachment in Postwar America," *British Journal for the History of Science* 44, no. 3 (2011): 401–26; "'The Father of Ethology and the Foster Mother of Ducks': Konrad Lorenz as an Expert on Motherhood," *ISIS* 100, no. 2 (2009): 263–91; and "Mothers, Machines, and Morals: Harry Harlow's Work on Primate Love from Lab to Legend," *Journal of the History of the Behavioral Sciences* 45, no. 3 (2009): 193–218.

Notes

Introduction

1. For a critical analysis, see Diane E. Eyer, *Mother-Infant Bonding: A Scientific Fiction* (New Haven, CT: Yale University Press, 1992); see also Diane E. Eyer, *Motherguilt: How Our Culture Blames Mothers for What's Wrong with Society* (New York: Random House, 1996) for the continuing impact of attachment on pediatric and child rearing practices.

2. Sarah Blaffer Hrdy, *Mother Nature: Maternal Instincts and the Shaping of the Species* (London: Vintage, 2000); Martha C. Nussbaum, *Upheavals of Thought: The Intelligence of Emotions* (Cambridge: Cambridge University Press, 2001).

3. John Bowlby, "Maternal Care and Mental Health," *Bulletin of the World Health Organization* 3 (1951): 355–534; Bowlby, *Child Care and the Growth of Love,* ed. Margery Fry (Harmondsworth, UK: Penguin, 1953).

4. John Bowlby, "The Nature of the Child's Tie to His Mother," *International Journal of Psychoanalysis* 39 (1958): 350–73; Bowlby, *Attachment*, vol. 1 of *Attachment and Loss* (New York: Basic Books, 1969).

5. See, for example, Irenäus Eibl-Eibesfeldt, *Love and Hate: The Natural History of Behavior Patterns* (New York: Holt, Rinehart and Winston, 1972), chap. 10; Hrdy, *Mother Nature*; Elaine Morgan, *The Descent of the Child: Human Evolution from a New Perspective* (London: Penguin, 1996), 117; Susan Allport, *A Natural History of Parenting: A Naturalist Looks at Parenting in the Animal World and Ours* (New York: Three Rivers Press, 1998), 167; and Thomas Lewis, Fari Amini, and Richard Lannon, *A General Theory of Love* (New York: Vintage, 2000).

6. Hamilton Cravens, *The Triumph of Evolution: The Heredity-Environment Controversy, 1900–1941* (Baltimore: Johns Hopkins University Press, 1988); Carl Degler, *In Search of Human Nature: The Decline and Revival of Darwinism in American Social Thought* (New York: Oxford University Press, 1991); Nancy Cott, *The Grounding of Modern Feminism* (New Haven, CT: Yale University Press, 1987); Rima D. Apple, *Perfect Motherhood: Science and Childrearing in America* (New Brunswick, NJ: Rutgers University Press, 2006).

7. Jan Lewis, "Mother's Love: The Construction of an Emotion in Nineteenth-Century America," in *Mothers and Motherhood: Readings in American History*, ed. Rima D. Apple and Janet Golden (Columbus: Ohio State University Press, 1997), 59; Ruth Bloch, "American Feminine Ideals in Transition: The Rise of the Moral Mother, 1785–1815," *Feminist Studies* 4 (1978): 100–126; Peter N. Stearns and Carol Z. Stearns, "Emotionology: Clarifying the History of Emotions and Emotional Standards," *American Historical Review* 90 (1985): 813–36.

8. Kathleen W. Jones, *Taming the Troublesome Child: American Families, Child Guidance, and the Limits of Psychiatric Authority* (Cambridge, MA: Harvard University Press, 1999), 2.

9. Charles E. Rosenberg and Carroll Smith-Rosenberg, "The Female Animal: Medical and Biological Views of Women," in Charles Rosenberg, *No Other Gods: On Science and American Social Thought* (Baltimore: Johns Hopkins University Press, 1976), 54–70; Cynthia Eagle Russett, *Sexual Science: The Victorian Construction of Womanhood* (Cambridge, MA: Harvard University Press, 1989).

10. On the history of views about maternal instincts, see Elisabeth Badinter, *Mother Love: Myth and Reality* (New York: Macmillan, 1981); Russett, *Sexual Science;* Stephanie A. Shields, "To Pet, Coddle, and 'Do For': Caretaking and the Concept of Maternal Instinct," in *In the Shadow of the Past: Psychology Portrays the Sexes*, ed. Miriam Lewin (New York: Columbia University Press, 1984), 256–73; and Marga Vicedo, "Mother Love and Human Nature: A History of the Maternal Instinct" (PhD diss., Harvard University, 2005). On early views of mother love, see Bloch, "American Feminine Ideals in Transition," and Lewis, "Mother's Love." For historical studies of the ideology of blaming mothers, see Molly Ladd-Taylor and Lauri Umansky, eds., *"Bad" Mothers: The Politics of Blame in Twentieth-Century America* (New York: New York University Press, 1998), and Jones, *Taming the Troublesome Child.* On the role of science in shaping views about motherhood, see Apple, *Perfect Motherhood.*

11. Rebecca Jo Plant, *Mom: The Transformation of Motherhood in Modern America* (Chicago: University of Chicago Press, 2010).

12. Bowlby, *Attachment,* 166. On different ways of appealing to the authority of nature, see Lorraine Daston and Fernando Vidal, eds., *The Moral Authority of Nature* (Chicago: University of Chicago Press, 2004).

13. The literature on the history of emotions is enormous. For influential reviews, see Stearns and Stearns, "Emotionology"; William M. Reddy, "Historical Research on the Self and Emotions," *Emotion Review* 1 (2009): 302–15; and Barbara H. Rosenwein, "Worrying about Emotions in History: Review Essay," *American Historical Review* 107 (2002): 821–45. Some recent studies are Peter N. Stearns and Jan Lewis, eds., *An Emotional History of the United States* (New York: New York University Press, 1998); Daniel M. Gross, *The Secret History of Emotion: From Aristotle's Rhetoric to Modern Brain Science* (Chicago: University of Chicago Press, 2006); Arlie Russell Hochschild, *The Managed Heart* (Berkeley: Uni-

versity of California Press, 2003); and Thomas Dixon, *From Passions to Emotions: The Creation of a Secular Psychological Category* (Cambridge: Cambridge University Press, 2003). For the literature on mother love, see note 10 above and references in the chapters.

Chapter One

1. Kathleen W. Jones, *Taming the Troublesome Child: American Families, Child Guidance, and the Limits of Psychiatric Authority* (Cambridge, MA: Harvard University Press, 1999).

2. Mari Jo Buhle, *Feminism and Its Discontents: A Century of Struggle with Psychoanalysis* (Cambridge, MA: Harvard University Press, 1998); Rebecca Jo Plant, *Mom: The Transformation of Motherhood in Modern America* (Chicago: University of Chicago Press, 2010).

3. The number three million comes from Edward A. Strecker, *Their Mothers' Sons: The Psychiatrist Examines an American Problem* (Philadelphia: Lippincott, 1946), 18. Ellen Herman notes that 550,000 servicemen were discharged on neuropsychiatric grounds, and the Selective Service examiners rejected about 1,846,000 for the same reason. See Ellen Herman, *The Romance of American Psychology: Political Culture in the Age of Experts* (Berkeley: University of California Press, 1995), 88–89.

4. On concerns about the machine, see John M. Jordan, *Machine-Age Ideology: Social Engineering and American Liberalism, 1911–1939* (Chapel Hill: University of North Carolina Press, 1994); Paul N. Edwards, *The Closed World: Computers and the Politics of Discourse in Cold War America* (Cambridge, MA: MIT Press, 1996).

5. Edward A. Strecker, "Presidential Address," *American Journal of Psychiatry* 101 (1944): 3.

6. Gordon W. Allport, "Scientific Models and Human Morals," *Psychological Review* 54 (1947): 182, 189.

7. "Psychology Group Issues Manifesto Against War," *Science News Letter* 32 (November 13, 1937): 307; John M. Fletcher, "The Verdict of Psychologists on War Instincts," *Scientific Monthly* 35 (1932): 142. See J. R. Rees, "What War Taught Us about Human Nature," *New York Times Magazine* 11 (March 17, 1946): 54–55. J. S. Huxley, "Is War Instinctive—And Inevitable?" *New York Times Magazine*, February 10, 1946, 7 ff., and James G. Needham, "Wars Caused by Our Instincts," *Science Digest* 11 (March 1942): 1–4.

8. William James, *Principles of Psychology* (1890; reprint, New York: Henry Holt, 1900), 442; J. B. Watson, *Psychology from the Standpoint of a Behaviorist* (Philadelphia: Lippincott, 1919), 231; William McDougall, *An Introduction to Social Psychology* (1908; reprint, Boston: Luce, 1916), chap. 3.

9. Albert Deutsch, "World Mental Health Congress to Tackle Causes of War and Peace," *New York Times*, November 7, 1947; Catherine Mackenzie, "World Parley Set on Mental Health," *New York Times*, November 4, 1947. For the rise of guidance clinics, child science, and social agencies, see Joseph M. Hawes, *Children between the Wars: American Childhood, 1920–1940* (New York: Twayne, 1997). See also Steven Mintz and Susan Kellogg, *Domestic Revolutions: A Social History of American Family Life* (New York: Free Press, 1988). On 165, Mintz and Kellogg write: "Although it is uncertain whether juvenile delinquency, parental neglect, or child abuse represented greater problems in the United States in the 1940s than in the 1930s, there is a mass of evidence indicating that social workers, psychologists, and public leaders were deeply troubled by the impact of the war on the nation's young people." United Nations General Assembly Resolution 217 A (III), December 10, 1948; quotation from Article 25 (2).

10. George K. Pratt, *Soldier to Civilian: Problems of Readjustment* (New York: McGraw-Hill, 1944). For a study by one of the psychoanalysts discussed here, see Therese Benedek, *Insight and Personality Adjustment: A Study of the Psychological Effects of War* (New York: Ronald Press, 1946).

11. Plant, *Mom*, 104.

12. Nancy F. Cott, *Public Vows: A History of Marriage and the Nation* (Cambridge, MA: Harvard University Press, 2000), 189–90. See also Mintz and Kellogg, *Domestic Revolutions*, 170–71. On postwar families, see Elaine Tyler May, *Homeward Bound: American Families in the Cold War Era* (New York: Basic Books, 1988), and Jessica Weiss, *To Have and to Hold: Marriage, the Baby Boom, and Social Change* (Chicago: University of Chicago Press, 2000).

13. Cott, *Public Vows*, 190; Susan M. Hartmann, "Prescriptions for Penelope: Literature on Women's Obligations to Returning World War II Veterans," *Women's Studies* 5 (1978): 223–39.

14. Mirra Komarovsky, "Cultural Contradictions and Sex Roles," *American Journal of Sociology* 52 (1946): 184, 185. See also Komarovsky, "Functional Analysis of Sex Roles," *American Sociological Review* 15 (1950): 508–16; *Life* article cited in William H. Chafe, *The Paradox of Change: American Women in the 20th Century* (New York: Oxford University Press, 1991), 175.

15. Ford Foundation, *Report of the Study for the Ford Foundation on Policy and Program* (Detroit: Ford Foundation, 1949), 44.

16. William C. Menninger, *Psychiatry in a Troubled World: Yesterday's War and Today's Challenge* (New York: Macmillan, 1948); Richard Pells, *The Liberal Mind in a Conservative Age: American Intellectuals in the 1940s and 1950s* (Middletown, CT: Wesleyan University Press, 1985). For the rise of psychology as a profession and a creed after World War II, see Herman, *Romance of American Psychology*, and James H. Capshew, *Psychologists on the March: Science, Practice, and Professional Identity in America, 1929–1969* (Cambridge, MA: Cambridge University Press, 1999). For the history of psychiatry, see Gerald N. Grob, *The Mad among Us:*

A History of the Care of America's Mentally Ill (Cambridge, MA: Harvard University Press, 1994), and Elizabeth Lunbeck, *The Psychiatric Persuasion: Knowledge, Gender, and Power in Modern America* (Princeton, NJ: Princeton University Press, 1994). On the passage of the Mental Health Act, see Grob, *Mad among Us*, chap. 8 (numbers from Grob, *Mad among Us*, 218).

17. Franz Alexander, *Our Age of Unreason: A Study of the Irrational Forces in Social Life*, rev. ed. (1942; Philadelphia: Lippincott, 1951), 235.

18. John Bowlby, "Psychology and Democracy," *Political Quarterly* 17 (1946): 76.

19. Helen Witmer, "Introduction," in *Symposium on the Healthy Personality*, ed. Milton J. E. Senn (New York: Josiah Macy Jr. Foundation, 1950), 13. See also Helen Leland Witmer and Ruth Kotinsky, eds., *Personality in the Making: The Fact-Finding Report of the Midcentury White House Conference on Children and Youth* (New York: Harper & Row, 1952).

20. Robert Briffault, *The Mothers: A Study of the Origins of Sentiments and Institutions*, 3 vols. (New York: Macmillan, 1927). For the gendered history of emotions, see Stephanie A. Shields, *Speaking from the Heart: Gender and the Social Meaning of Emotion* (Cambridge: Cambridge University Press, 2002). On the maternal instinct, see Cynthia Eagle Russett, *Sexual Science: The Victorian Construction of Womanhood* (Cambridge, MA: Harvard University Press, 1989); Stephanie A. Shields, "To Pet, Coddle, and 'Do For': Caretaking and the Concept of the Maternal Instinct," in *In the Shadow of the Past: Psychology Portrays the Sexes*, ed. Miriam Lewin (New York: Columbia University Press, 1984), 256–73; Marga Vicedo, "Mother Love and Human Nature: A History of the Maternal Instinct" (PhD diss., Harvard University, 2005).

21. Buhle, *Feminism and Its Discontents*, 123.

22. Spitz to Anna Freud, August 4, 1947, box 96, folder 7 ("Spitz, René A., 1946–74"), Anna Freud Papers, Sigmund Freud Collection, Manuscript Division, Library of Congress, Washington, DC.

23. Sigmund Freud, "On Narcissism: An Introduction" (1914), in *The Standard Edition of the Complete Psychological Works of Sigmund Freud*, trans. under the general editorship of James Strachey in collaboration with Anna Freud and assisted by Alix Strachey and Alan Tyson, 24 vols. (London: Hogarth, 1953–74), 14:78.

24. Anna Freud, *The Ego and the Mechanisms of Defense* (London: Hogarth, 1936). On Anna Freud, see Janet Sayers, *Mothers of Psychoanalysis: Helene Deutsch, Karen Horney, Anna Freud, and Melanie Klein* (New York: Norton 1991); Lisa Appignanesi and John Forrester, *Freud's Women* (London: Phoenix, 2005), and Elisabeth Young-Bruehl, *Anna Freud: A Biography*, 2nd ed. (New Haven, CT: Yale University Press, 2008).

25. Anna Freud and Dorothy Burlingham, *Infants without Families: Reports on the Hampstead Nurseries* (1943), in vol. 3 of *The Writings of Anna Freud* (New York: International Universities Press, 1973), 129. A short version of the reports

on the war children was published in the United States. See Anna Freud and Dorothy T. Burlingham, *War and Children: A Message to American Parents*, 2nd ed. (New York: International University Press, 1944). See also Anna Freud and Sophie Dann, "An Experiment in Group Upbringing," *Psychoanalytic Study of the Child* 6 (1951): 127–68.

26. Freud and Burlingham, *Infants without Families*, 129, 130.

27. On Anna Freud's influence, see Sayers, *Mothers of Psychoanalysis*, 158.

28. On Levy, see "David M. Levy, M.D.," *American Journal of Orthopsychiatry* 8 (1938): 769–70.

29. David M. Levy, "Primary Affect Hunger," *American Journal of Psychiatry* 94 (1937): 643–44. See "Personality Vitamin," *Science News Letter* 31 (May 22, 1937): 326.

30. Margarethe A. Ribble, "Disorganizing Factors of Infant Personality," *American Journal of Psychiatry* 98 (1941): 459, 460. Ribble sometimes used "Margarethe," and other times she used "Margaret." In what follows, I will use "Margaret" unless I am quoting.

31. Margaret Ribble, *The Rights of Infants: Early Psychological Needs and Their Satisfaction* (New York: Columbia University Press, 1943), viii, ix.

32. Ribble, *Rights of Infants*, viii, xix, 8, 13.

33. Ibid., 9, 8, 14.

34. See William G. Bach, "The Influence of Psychoanalytic Thought on Benjamin Spock's *Baby and Child Care*," *Journal of the History of the Behavioral Sciences* 10 (1974): 91–94; Lynn Z. Bloom, *Doctor Spock: Biography of a Conservative Radical* (Indianapolis, IN: Bobbs-Merrill, 1972). Also see Thomas Maier, *Dr. Spock: An American Life* (New York: Harcourt Brace, 1998); A. Michael Sulman, "The Humanization of the American Child: Benjamin Spock as a Popularizer of Psychoanalytic Thought," *Journal of the History of the Behavioral Sciences* 9 (1973): 258–65; and Michael Zuckerman, "Dr. Spock: The Confidence Man," in *The Family in History*, ed. Charles E. Rosenberg (Philadelphia: University of Pennsylvania Press, 1975), 179–207.

35. Benjamin Spock and Michael B. Rothenberg, *Dr. Spock's Baby and Child Care* (1945; reprint, New York: E. P. Dutton, 1985), 3.

36. On Spitz, see Herbert S. Gaskill, "In Memoriam: René A. Spitz, 1887–1974," *Psychoanalytic Study of the Child* 31 (1976): 1–3.

37. René Spitz, *Grief: A Peril in Infancy* (1947), film available at the Archives of the History of American Psychology and the US National Library of Medicine, Bethesda, MD.

38. René A. Spitz, "Hospitalism: An Inquiry into the Genesis of Psychiatric Conditions in Early Childhood," *Psychoanalytic Study of the Child* 1 (1945): 53–74; René A. Spitz and Katherine M. Wolf, "Hospitalism: A Follow-up Report, *Psychoanalytic Study of the Child* 2 (1946): 113–17; Spitz, "Anaclitic Depression: An Inquiry into the Genesis of Psychiatric Conditions in Early Childhood, II," *Psy-*

choanalytic Study of the Child 2 (1946): 313–42; Spitz, "The Importance of the Mother-Child Relationship during the First Year of Life," *Mental Health Today* 7 (1948): 7–13; Spitz, "The Role of Ecological Factors in Emotional Development in Infancy," *Child Development* 20 (1949): 145–55.

39. Floyd M. Crandall, "Hospitalism," *Archives of Pediatrics* 14 (1897): 452.

40. Lucy Freeman, "Emotions of Baby Held First to Gain," *New York Times*, May 14, 1949, 10.

41. Therese Benedek, "The Psychosomatic Implications of the Primary Unit: Mother-Child," *American Journal of Orthopsychiatry* 19 (1949): 649, 651. On Benedek, see Thomas G. Benedek, "A Psychoanalytic Career Begins: Therese F. Benedek, M.D.—A Documentary Biography," *Annual of Psychoanalysis* 7 (1979): 3–15; and Doris Weidemann, *Leben und Werk von Therese Benedek, 1892–1977: Weibliche Sexualität und Psychologie des Weiblichen* (Frankfurt: Peter Lang, 1988).

42. Karl A. Menninger, with the collaboration of Jeanetta Lyle Menninger, *Love against Hate* (New York: Harcourt, Brace, 1942), 32.

43. Ribble, *Rights of Infants*, 42.

44. David M. Levy, *Maternal Overprotection* (1943; reprint, New York: Norton, 1966).

45. David M. Levy, "Psychopathic Personality and Crime," *Journal of Educational Sociology* 16 (1942): 100, 106.

46. These reviews, among many others, can be found in box 13.35 and box 13.9, David Levy Papers, Oskar Diethelm Library, DeWitt Wallace Institute for the History of Psychiatry, Weill Cornell Medical College, New York. Montagu's was published in *Psychiatry*, Reed's in *American Catholic Sociological Review*, and Sears's in *American Journal of Psychology*.

47. Review in *Medical Record*, box 13.9, Levy Papers.

48. Philip Wylie, *Generation of Vipers* (New York: Rinehart, 1942), 197. On Wylie, see Truman Frederick Keefer, *Philip Wylie* (Boston: Twayne, 1977), and Robert Howard Barshay, *Philip Wylie: The Man and His Work* (New York: University Press of America, 1979). For an excellent discussion of Wylie's views and the response to them in the context of changing views about mothers in the postwar era, see Plant, *Mom*. For a discussion of momism in the context of psychoanalytic views about child rearing and their relation to studies of national character, see Buhle, *Feminism and Its Discontents*, chap. 4. On momism, see also Ruth Feldstein, *Motherhood in Black and White: Race and Sex in American Liberalism, 1930–1965* (Ithaca, NY: Cornell University Press, 2000), and Jennifer Terry, "'Momism' and the Making of Treasonous Homosexuals," in *"Bad" Mothers: The Politics of Blame in Twentieth-Century America*, ed. Molly Ladd-Taylor and Lauri Umansky (New York: New York University Press, 1998), 169–90.

49. Buhle, *Feminism and Its Discontents*, 127.

50. Karl A. Menninger to Wylie, April 9, 1943, box 232, folder 2, Philip Wylie

Papers, Manuscripts Division, Department of Rare Books and Special Collections, Princeton University Library, Princeton, NJ.

51. Strecker, *Their Mothers' Sons*, 5. A long book summary, "What's Wrong with American Mothers?" was published in the *Saturday Evening Post*, October 26, 1946, 14–15, 85–99, 102, 104. See also Edward A. Strecker, "Motherhood and Momism—Effect on the Nation," *University of Western Ontario Medical Journal* 16 (1946): 59–77. On Strecker, see Lauren H. Smith, "Edward A. Strecker, M.D.: A Biographical Sketch," *American Journal of Psychiatry* 101 (July 1944): 9–11.

52. Strecker, *Their Mothers' Sons*, 13.

53. Ibid., 6, 21.

54. Ibid., 23, 30. See especially chap. 10 ("Mom in a Bottle"), 122–27, chap. 11 ("Homosexuality"), 128–32, and chap. 5 ("Mom Types"), 54–69.

55. Ibid., 148, 133.

56. Frieda Fromm-Reichman, "Notes on the Development of Treatment of Schizophrenics by Psychoanalytic Psychotherapy," *Psychiatry* 11 (1948): 265; Leo Kanner, "Autistic Disturbances of Affective Contact," *Nervous Child* 2 (1943): 217–50. Originally, Bettelheim did not favor selecting the mother as the cause of childhood schizophrenia, as can be seen in Bruno Bettelheim, "Schizophrenia as a Reaction to Extreme Situations," *American Journal of Orthopsychiatry* 26 (1956): 507–18. But he moved in that direction, and his work became the most visible representative of this position. See Bettelheim, *The Empty Fortress: Infantile Autism and the Birth of the Self* (New York: Free Press, 1967).

57. Erik H. Erikson, *Childhood and Society* (New York: Norton, 1950).

58. See Ellen Herman, "Families Made by Science: Arnold Gesell and the Technologies of Modern Child Adoption," *ISIS* 92 (2001): 684–715.

59. Erik Erikson, "Growth and Crises of the 'Healthy Personality,'" in *Personality in Nature, Society and Culture*, ed. Clyde Kluckhohn and Henry A. Murray (New York: Knopf, 1953), 195.

60. Erikson, *Childhood and Society*, 249, 250.

61. Ibid., 254, 250.

62. On Erikson's life, see Lawrence Friedman, *Identity's Architect: A Biography of Erik H. Erikson* (New York: Scribner, 1999). On his name change, see 143–47.

63. On Bowlby, see Jeremy Holmes, *John Bowlby and Attachment Theory* (New York: Routledge, 1993), and Suzan Van Dijken, *John Bowlby: His Early Life; a Biographical Journey into the Roots of Attachment Theory* (London: Free Association Books, 1998).

64. Bowlby's travel books can be found in the John Bowlby collection at the Wellcome Library, London. See PP/BOW/D.4/8, "WHO 1950 Notebooks." For the United States travel, see also PP/BOW/B.1/12 "Mar 1950. Boston, New York, Chicago."

65. William Goldfarb, "Infant Rearing and Problem Behavior," *American Journal of Orthopsychiatry* 13 (1943): 250; Goldfarb, "Psychological Privation in

Infancy and Subsequent Adjustment," *American Journal of Orthopsychiatry* 15 (1945): 252.

66. John Bowlby, "Maternal Care and Mental Health," *Bulletin of the World Health Organization* 3 (1951): 355–534.

67. Bowlby, "Maternal Care and Mental Health," 59, emphasis added; 13.

68. John Bowlby, "The Influence of Early Environment in the Development of Neurosis and Neurotic-Character," *International Journal of Psychoanalysis* 21 (1940): 154, 155.

69. Bowlby, "Influence of Early Environment," 156, 162, 163, 158.

70. Ibid., 163, 164.

71. John Bowlby, "Forty-Four Juvenile Thieves: Their Characters and Home-Life," *International Journal of Psychoanalysis* 25 (1944): 113.

72. Bowlby, "Influence of Early Environment," 164.

73. Bowlby, "Maternal Care and Mental Health," 67.

74. Ibid., 158.

75. Ibid., 53.

76. Ibid., 46, 61.

77. Inge Bretherton, "The Origins of Attachment Theory: John Bowlby and Mary Ainsworth," *Developmental Psychology* 28 (1992): 761.

78. John Bowlby, *Child Care and the Growth of Love*, ed. Margery Fry (Harmondsworth, UK: Pelican, 1953), 15. Reference to Lorenz is on 183.

79. Ribble, *Rights of Infants*, viii.

Chapter Two

1. John Bowlby, *Child Care and the Growth of Love*, ed. Margery Fry (Harmondsworth, UK: Pelican, 1953), 15. Bowlby included "Lorenz, K." in the "List of Authorities referred to but not named" on p. 183.

2. The most complete history of ethology is Richard W. Burkhardt, *Patterns of Behavior: Konrad Lorenz, Niko Tinbergen, and the Founding of Ethology* (Chicago: University of Chicago Press, 2005); see also John R. Durant, "Innate Character in Animals and Man: A Perspective on the Origins of Ethology," in *Biology, Medicine and Society, 1840–1940*, ed. Charles Webster (Cambridge: Cambridge University Press, 1981), 157–92; Donald A. Dewsbury, *Comparative Psychology in the Twentieth Century* (Stroudsburg, PA: Hutchinson Ross, 1984); Richard W. Burkhardt Jr., "The Founders of Ethology and the Problem of Human Aggression: A Study in Ethologists' Ecologies," in *The Animal/Human Boundary: Historical Perspectives*, ed. Angela N. H. Creager and William Chester Jordan (Rochester, NY: University of Rochester Press, 2002), 265–304. For the other authors, see further notes in this chapter.

3. On Lorenz's life and work, see Klaus Taschwer and Benedikt Föger, *Konrad*

Lorenz: Biographie (Vienna: Paul Zsolnay, 2003). See also Burkhardt, *Patterns of Behavior*.

4. On Heinroth, see Burkhardt, *Patterns of Behavior*, 135ff., and Karl Bühler, *Die Krise der Psychologie* (Jena: Gustav Fischer, 1927). On Bühler, see James F. T. Bugental, ed., "Symposium on Karl Bühler's Contributions to Psychology," *Journal of General Psychology* 75, no. 2 (1966): 181–219. On his influence on Lorenz, see Veronika Hofer, "Konrad Lorenz als Schüler von Karl Bühler: Diskussion der neu entdeckten Quellen zu den persönlichen und inhaltlichen Positionen zwischen Karl Bühler, Konrad Lorenz und Egon Brunswik," *Die Zeitgeschichte* 28 (2001): 135–59.

5. Lorenz to Heinroth, February 22, 1931; cited in Burkhardt, *Patterns of Behavior*, 141.

6. See Konrad Lorenz, "Beobachtungen an Dohlen," *Journal für Ornithologie* 75 (1927): 511–19; Lorenz, "Beiträge zur Ethologie sozialer Corviden," *Journal für Ornithologie* 79 (1931): 67–127; Lorenz, "Betrachtungen über das Erkennen der arteigenen Triebhandlungen der Vögel," *Journal für Ornithologie* 80 (1932): 50–98; Lorenz, "Beobachtungen an freifliegenden zahmgehaltenen Nachtreihern," *Journal für Ornithologie* 82 (1934): 160–61; and Lorenz, "Über die Bildung des Instinktbegriffes," *Die Naturwissenschaften* 25 (1937): 289–300, 307–18, 324–31.

7. Konrad Lorenz, "Der Kumpan in der Umwelt des Vogels, der Artgenosse als auslösendes Moment sozialer Verhaltensweisen," *Journal für Ornithologie* 83 (1935): 137–215, 289–413. This article was published in three English translations, but only the last one is complete: Lorenz, "The Companion in the Bird's World," *Auk* 54 (1937): 245–73; Lorenz, "Companionship in Bird Life: Fellow Members of the Species as Releasers of Social Behavior," in *Instinctive Behavior: The Development of a Modern Concept*, ed. Claire H. Schiller (New York: International Universities Press, 1957), 83–128; and Lorenz, "Companions as Factors in the Bird's Environment: The Conspecific as the Eliciting Factor for Social Behaviour Patterns," in Lorenz, *Studies in Animal and Human Behavior*, vol. 1 (Cambridge, MA: Harvard University Press, 1970), 101–258 (this version will be cited here).

8. Lorenz borrowed the concepts of "companion" and "releaser" from the work of Jakob von Uexküll, who had incorporated some of Lorenz's observations on the social behavior of jackdaws into his own theory of the perceptual worlds of animals. See Jacob von Uexküll, "A Stroll through the Worlds of Animals and Men: A Picture Book of Invisible Worlds," in Schiller, *Instinctive Behavior*, 5–80. On Uexküll see Anne Harrington, *Reenchanted Science: Holism in German Culture from Wilhelm II to Hitler* (Princeton, NJ: Princeton University Press, 1996), and see Burkhardt, *Patterns of Behavior*, on early influences on Lorenz's work.

9. Lorenz, "Companions as Factors," 254, 258.

10. On Tinbergen, see Hans Kruuk, *Niko's Nature: A Life of Niko Tinbergen and His Science of Animal Behavior* (Oxford: Oxford University Press, 2003); Burkhardt, *Patterns of Behavior*; D. R. Röell, *The World of Instinct: Niko Tinber-*

gen and the Rise of Ethology in the Netherlands (1920–1950) (Assen, Netherlands: Van Gorcum, 2000). For Tinbergen's reflections on his career, see Niko Tinbergen, "Watching and Wondering," in *Studying Animal Behavior: Autobiographies of the Founders*, ed. Donald A. Dewsbury (Chicago: University of Chicago Press, 1989), 431–63. On the early collaborative work between Lorenz and Tinbergen in 1936–1937, see Burkhardt, *Patterns of Behavior*, 187–88, 205–13.

11. Konrad Lorenz and Niko Tinbergen, "Taxis and Instinctive Behaviour Pattern in Egg-Rolling by the Greylag Goose," in Lorenz, *Studies in Animal and Human Behaviour*, 1:316–50; originally published as "Taxis und Instinkthandlung in der Eirollbewegung der Graugans," *Zeitschrift für Tierpsychologie* 2 (1938): 1–29.

12. Burkhardt, *Patterns of Behavior*, passim.

13. On the role of film in ethology, see Gregg Mitman, *Reel Nature: America's Romance with Wildlife on Film* (Cambridge, MA: Harvard University Press, 1999), chap. 3; Tania Munz, "Die Ethologie des wissenschaftlichen Cineasten: Karl von Frisch, Konrad Lorenz und das Verhalten der Tiere im Film," *montage AV* 14, no. 4 (2005): 52–68.

14. Mitchell G. Ash, "Psychology and Politics in Interwar Vienna: The Vienna Psychological Institute, 1922–1942," in *Psychology in Twentieth-Century Thought and Society*, ed. M. G. Ash and W. R. Woodward (Cambridge: Cambridge University Press, 1987), 157.

15. On Lorenz and National Socialism, see Burkhardt, *Patterns of Behavior*, chap. 5; Ute Deichmann, *Biologists under Hitler* (Cambridge, MA: Harvard University Press, 1996), 179–205; Benedikt Föger and Klaus Taschwer, *Die andere Seite des Spiegels: Konrad Lorenz und der Nationalsozialismus* (Vienna: Czernin, 2001); and Theodora Kalikow, "Konrad Lorenz's Ethological Theory, 1939–1943: 'Explanations' of Human Thinking, Feeling, and Behaviour," *Philosophy of the Social Sciences* 6 (1976): 15–34.

16. Konrad Lorenz, "Über Ausfallserscheinungen im Instinktverhalten von Haustieren und ihre sozialpsychologische Bedeutung," in *Charakter und Erziehung: Bericht über den 16. Kongress der deutschen Gesellschaft für Psychologie in Bayreuth*, ed. Otto Klemm (Leipzig: Johann Ambrosius Barth, 1939), 139–47; Konrad Lorenz, "Durch Domestikation verursachte Störungen arteigenen Verhaltens," *Zeitschrift für angewandte Psychologie und Charakterkunde* 59, nos. 1–2 (1940): 2–81.

17. Konrad Lorenz in *Königsberger Allgemeine Zeitung*, November 2, 1940. Cited in Föger und Taschwer, *Die andere Seite*, 134–35.

18. Konrad Lorenz, "Die angeborenen Formen möglicher Erfahrung," *Zeitschrift für Tierpsychologie* 5 (1943): 235–409.

19. Konrad Lorenz, *King Solomon's Ring: New Light on Animal Ways* (New York: Crowell, 1952); originally published as *Er redete mit dem Vieh, den Vögeln und den Fischen* (Vienna: Borotha-Schoeler, 1949). In some editions the English

translation included a subtitle: *He Spoke with the Beasts, the Birds, and the Fish.* According to Tania Munz, the book went through some forty printings and was translated into twelve languages. See Tania Munz, "'My Goose Child Martina': The Multiple Uses of Geese in Konrad Lorenz's Writings on Animals, 1935–1988," *Historical Studies in the Natural Sciences* 41 (2011): 431.

20. Konrad Lorenz, "The Comparative Method in Studying Innate Behaviour Patterns," *Symposia of the Society for Experimental Biology* 4 (1950): 221, 222, 234.

21. Ibid., 265.

22. Niko Tinbergen, *The Study of Instinct* (Oxford: Clarendon Press, 1951), 1, 2.

23. On Craig's work and on his influence on Lorenz, see Burkhardt, *Patterns of Behavior*, chaps. 1 and 3, respectively.

24. Lorenz, "Companions as Factors," 250–51. On 250–52 Lorenz presents a short history of the concept of instinct, noting the sources of some of the elements he used for his own conception. See also Wallace Craig, "Appetites and Aversions as Constituents of Instincts," *Biological Bulletin* 34 (1918): 91–107. On Lorenz's views about instincts, see Burkhardt, *Patterns of Behavior*; Theodora J. Kalikow, "History of Konrad Lorenz's Ethological Theory, 1927–1939: The Role of Metatheory, Theory, Anomaly and New Discoveries in a Scientific 'Evolution,'" *Studies in the History and Philosophy of Science* 6 (1975): 331–41; Robert Richards, "The Innate and the Learned: The Evolution of Konrad Lorenz's Theory of Instinct," *Philosophy of the Social Sciences* 4 (1974): 111–33; Colin G. Beer, "Darwin, Instinct, and Ethology," *Journal of the History of the Behavioral Sciences* 19 (1983): 68–80; and Ingo Brigandt, "The Instinct Concept of the Early Konrad Lorenz," *Journal of the History of Biology* 38 (2005): 571–608.

25. Konrad Lorenz, "The Establishment of the Instinct Concept" (1937), in Lorenz, *Studies in Animal and Human Behavior*, 1:259–315.

26. Lorenz, "Comparative Method," 255. Tinbergen's model of motivation was different from Lorenz's model. Tinbergen postulated a hierarchical system of centers. See Tinbergen, *Study of Instinct*, 125. Also see Lorenz, "The Past Twelve Years in the Comparative Study of Behavior" (1952), in Schiller, *Instinctive Behavior*, 288–310.

27. J. B. S. Haldane, "The Sources of Some Ethological Notions," *British Journal of Animal Behaviour* 4 (1956): 162–64. For a comparison of Lorenz's and MacDougall's models, see Paul E. Griffiths, "Instinct in the '50s: The British Reception of Konrad Lorenz's Theory of Instinctive Behavior," *Biology and Philosophy* 19 (2004): 609–31.

28. Lorenz, "Comparative Method," 221.

29. Lorenz, "A Consideration of Methods of Identification of Species-Specific Instinctive Behaviour Patterns in Birds" (1932), in Lorenz, *Studies in Animal and Human Behavior*, 1:57–100; Lorenz, "Establishment of the Instinct Concept."

30. Lorenz, "Consideration of Methods," 65; Lorenz, "Companions as Factors," 248.

31. See Burkhardt, *Patterns of Behavior*, chap. 3.

32. Lorenz's most extended discussion of deprivation experiments is found in "The Value and the Limitations of the Deprivation Experiment," chap. 7 of his *Evolution and Modification of Behavior* (Chicago: University of Chicago Press, 1965), 83–100.

33. Lorenz, "Companions," 126; emphasis in original.

34. Ibid., 245.

35. Ibid., 124ff.

36. Ibid., 168.

37. Ibid., 185.

38. Ibid., 244.

39. Lorenz, "Die angeborenen Formen," 274; my translation.

40. Ibid., 276.

41. Konrad Lorenz, "Companions," 251; Lorenz, "Comparative Method," 263; William James, *Principles of Psychology* (1890; reprint, New York: Holt, 1900), 442; John B. Watson, *Psychology from the Standpoint of a Behaviorist* (Philadelphia: Lippincott, 1919), 231; and William McDougall, *An Introduction to Social* Psychology (1908; reprint, Boston: Luce, 1916), 29.

42. Lorenz, "Comparative Method," 265.

43. Konrad Lorenz, "Über angeborene Instinktformeln beim Menschen," *Deutsche medizinische Wochenschrift* 78 (1953): 1566–69, 1600–1604.

44. The verbatim transcriptions of the proceedings of the study group meetings were published by the Tavistock Institute in 1956 (both the 1953 and the 1954 meetings), 1958, and 1960. They were later collected in a single volume: J. M. Tanner and Barbel Inhelder, eds., *Discussions on Child Development* (London: Tavistock, 1971), but with separate pagination for each original volume. All citations here refer to this edition.

45. Bowlby, in Tanner and Inhelder, *Discussions on Child Development*, 1:27.

46. Konrad Lorenz to Erwin Stresemann, May 14, 1934; quoted by Föger and Taschwer, *Andere Seite*, 55–56.

47. Konrad Lorenz, "Memorandum on Ethology," January 7, 1953, PP/BOW/H.132, John Bowlby Papers, Western Manuscripts and Archives, Wellcome Library, London; Lorenz, in Tanner and Inhelder, *Discussions on Child Development*, 1:117.

48. Lorenz, in Tanner and Inhelder, *Discussions on Child Development*, 1:215–16, quotation on 211.

49. Ibid., 222.

50. Ibid., 223.

51. See Tinbergen, *Study of Instinct*, 45. Tinbergen describes other examples of "supernormal" stimuli on 44–45.

52. Leta S. Hollingsworth, "Social Devices for Impelling Women to Bear and Rear Children," *American Journal of Sociology* 22 (1916): 19–29; Ruth Reed,

"Changing Conceptions of the Maternal Instinct," *Journal of Abnormal Psychology and Social Psychology* 18 (1923): 78–87. On the maternal instinct, see Marga Vicedo, "Mother Love and Human Nature: A History of the Maternal Instinct" (PhD diss., Harvard University, 2005).

53. Lorenz, in Tanner and Inhelder, *Discussions on Child Development*, 1: 227–28.

54. Mead, in ibid., 1:228.

55. Lorenz in Tanner and Inhelder, *Discussions on Child Development*, 3:36, 69; see also 45.

56. Mitman, *Reel Nature.*

57. "An Adopted Mother Goose: Filling a Parent's Role, a Scientist Studies Goslings' Behavior," *Life Magazine* 39 (July/August 1955): 73–74, 77–78. For an analysis of Lorenz's use of his role as "mother" to study animals, see Vicedo, "Outside or Inside the Animal?"

58. For an analysis of Lorenz's connection with greylag geese, see Klaus Taschwer, "Von Gänsen und Menschen: Über die Geschichte der Ethologie in Österreich und über ihren Protagonisten, den Forscher, Popularisator und Ökopolitiker Konrad Lorenz," in *Wissenschaft, Politik und Öffentlichkeit, von der Wiener Moderne bis zur Gegenwart,* ed. Mitchell G. Ash and Christian H. Stifter (Vienna: WUV, 2002), 331–51.

59. Munz, "'My Goose Child Martina.'"

60. Konrad Lorenz, *On Aggression* (1966; reprint, New York: Bantam, 1969), 203.

61. Munz, "'My Goose Child Martina,'" 432.

Chapter Three

1. On Bowlby, see references in chapter 1. On Bowlby's influence in the United States, see Barbara Ehrenreich and Deirdre English, *For Her Own Good: 150 Years of the Experts' Advice to Women* (New York: Anchor Books, 1978), 229; Mari Jo Buhle, *Feminism and Its Discontents: A Century of Struggle with Psychoanalysis* (Cambridge, MA: Harvard University Press, 1998), 162; Ann Hulbert, *Raising America: Experts, Parents, and a Century of Advice about Children* (New York: Vintage Books, 2003), 204–5; Christina Hardyment, *Dream Babies: Three Centuries of Good Advice on Child Care* (New York: Harper and Row, 1983), 236; Elaine Morgan, *The Descent of the Child: Human Evolution from a New Perspective* (London: Penguin, 1996), 118; Carl N. Degler, *At Odds: Women and the Family in America from the Revolution to the Present* (New York: Oxford University Press, 1980), 471; Julia Grant, *Raising Baby by the Book: The Education of American Mothers* (New Haven, CT: Yale University Press, 1998), 211; Maxine L. Margolis, *Mothers and Such: Views of American Women and Why They Changed* (Berkeley: University of California Press, 1984), 70; Molly Ladd-Taylor and Lauri

Umansky, introduction to *"Bad" Mothers: The Politics of Blame in Twentieth-Century America*, ed. Molly Ladd-Taylor and Lauri Umansky (New York: New York University Press, 1998), 1–28, esp. 14; John Byng-Hall, "An Appreciation of John Bowlby: His Significance for Family Therapy," *Journal of Family Therapy* 13 (1991): 5–16; Barbara Melosh, *Strangers and Kin: The American Way of Adoption* (Cambridge, MA: Harvard University Press, 2002), 75, credits Bowlby with having influenced the change in standards of placement by adoption agencies. Melosh notes the appearance of Bowlby's report in 1951 and also the fact that by 1955 "half of all agencies were placing children less than a month old." It is not clear whether Melosh thinks Bowlby's report was the main cause of that change or just one factor among others. The trend in child adoption had gone from two years to one year to six months. A contemporary study documented the shift toward adopting younger children, based on the data provided in the 1952 Child Welfare League of America report, and discussed the influence of Bowlby. See J. Richard Wittenborn, assisted by Barbara Myers, *The Placement of Adoptive Children* (Springfield, IL: Charles C. Thomas, 1957), 4; and Ellen Herman, *Kinship by Design: A History of Adoption in the Modern United States* (Chicago: University of Chicago Press, 2008), 133.

2. Ruth Bloch, "American Feminine Ideals in Transition: The Rise of the Moral Mother, 1785–1815," *Feminist Studies* 4 (1978): 101–26; Jan Lewis, "Mother's Love: The Construction of an Emotion in Nineteenth-Century America," in *Mothers and Motherhood: Readings in American History*, ed. Rima D. Apple and Janet Golden (Columbus: Ohio State University Press, 1997), 52–71.

3. John Bowlby, *Child Care and the Growth of Love*, ed. Margery Fry (Harmondsworth, UK: Pelican, 1953), 18, 50; see also 34–35. This is an abridged version of John Bowlby, "Maternal Care and Mental Health," *Bulletin of the World Health Organization* 3 (1951): 355–534. On the history and significance of objectivity in science, see Lorraine Daston and Peter Galison, *Objectivity* (Cambridge, MA: MIT Press, 2007).

4. Bowlby, *Child Care*, 75–76.

5. Ibid., 76.

6. Ibid., 84.

7. Ibid., 181.

8. John Bowlby, "Mother Is the Whole World," *Home Companion* (London), 1952, 29–33. Newspaper clippings in PP/BOW/A.4/1, "Presscuttings April 1952–May 1953," and PP/BOW/A.4/2, "Press cuttings 1953–1954." John Bowlby Papers, Wellcome Library, London. This is only a small sample of the many reviews Bowlby collected.

9. Nancy Cott, *Public Vows: A History of Marriage and the Nation* (Cambridge, MA: Harvard University Press, 2000), 191; Steven Mintz and Susan Kellogg, *Domestic Revolutions: A Social History of American Family Life* (New York: Free Press, 1988). On day care, see also Mary Frances Berry, *The Politics of Parenthood:*

Child Care, Women's Rights, and the Myth of the Good Mother (Harmondsworth, UK: Penguin, 1994).

10. Elaine Tyler May, *Homeward Bound: American Families in the Cold War Era* (New York: Basic Books, 1988). See also Ruth Feldstein, *Motherhood in Black and White: Race and Sex in American Liberalism, 1930–1965* (Ithaca, NY: Cornell University Press, 2000).

11. Gaile McGregor, "Domestic Blitz: A Revisionist Theory of the Fifties," *American Studies* 34 (1993): 8.

12. Cott, *Public Vows*, 189. See also Mintz and Kellogg, *Domestic Revolutions*, 170–71.

13. On delinquency and family life during the Cold War, see James Gilbert, *A Cycle of Outrage: America's Reaction to the Juvenile Delinquent in the 1950s* (New York: Oxford University Press, 1986), and Mintz and Kellogg, *Domestic Revolutions*.

14. Alfred C. Kinsey et al., *Sexual Behavior in the Human Female* (Philadelphia: W. B. Saunders, 1953); Mirra Komarovsky, *Women in the Modern World: Their Education and Their Dilemmas* (Boston: Little, Brown, 1953); Ashley Montagu, *The Natural Superiority of Women* (New York: Macmillan, 1953); Simone de Beauvoir, *The Second Sex* (New York: Vintage Books, 1953).

15. Joanne Meyerowitz, "Beyond the Feminine Mystique: A Reassessment of Postwar Mass Culture, 1946–1958," *Journal of American History* 79 (1993): 1455–82; McGregor, "Domestic Blitz"; Joanne Meyerowitz, ed., *Not June Cleaver: Women and Gender in Postwar America, 1945–1960* (Philadelphia: Temple University Press, 1994). On mothers, see Wini Breines, "Domineering Mothers in the 1950s: Image and Reality," *Women's Studies International Forum* 8 (1985): 601–8; Margolis, *Mothers and Such*, chap. 3; and Mintz and Kellogg, *Domestic Revolutions*. On the rise of scientific authority in the realm of child rearing, see Rima D. Apple, *Perfect Motherhood: Science and Childrearing in America* (New Brunswick, NJ: Rutgers University Press, 2006), and Peter N. Stearns, *Anxious Parents: A History of Modern Childrearing in America* (New York: New York University Press, 2003).

16. Beauvoir, *Second Sex*, 570. See Therese Benedek, review of *The Second Sex*, by Simone de Beauvoir, *Psychoanalytic Quarterly* 22 (1953): 264–67; Marjorie Grene, review of *The Second Sex*, by Simone de Beauvoir, *New Republic* 128 (March 9, 1953): 22–23; and "A Senior Panel Takes Aim at *The Second Sex*," *Saturday Review* (February 21, 1953): 26–31, 41. The panel of reviewers included a psychiatrist, Karl Menninger; a writer, Philip Wylie; an educator, Ashley Montagu; a housewife, Phyllis McGinley; an anthropologist, Margaret Mead; and a public official, Olive R. Goldman.

17. Alva Myrdal and Viola Klein, *Women's Two Roles: Home and Work* (London: Routledge and Kegan Paul, 1956), 78, 83.

18. Sonya Michel, *Children's Interests/Mothers' Rights: The Shaping of Amer-*

ica's Child Care Policy (New Haven, CT: Yale University Press, 1999), 151; *Employed Mothers and Child Care*, Bulletin of the Women's Bureau 246 (Washington, DC: GPO, 1953), iii, 1.

19. Kathleen W. Jones, *Taming the Troublesome Child: American Families, Child Guidance, and the Limits of Psychiatric Authority* (Cambridge, MA: Harvard University Press, 1999); Ellen Herman, *The Romance of American Psychology: Political Culture in the Age of Experts* (Berkeley: University of California Press, 1995); Mark Solovey, *Shaky Foundations: The Politics-Patronage-Social-Science Nexus in Cold War America* (New Brunswick, NJ: Rutgers University Press, 2013).

20. See chapter 1.

21. James Robertson, application to the Institute of Psychoanalysis for training, dated April 26, 1950, box 86, folder 18 ("Robertson, James, 1950–53"), Anna Freud Papers, Sigmund Freud Collection, Manuscript Division, Library of Congress, Washington, DC.

22. Robertson to Anna Freud, February 5, 1953; Anna Freud to Robertson, February 20, 1953, box 86, folder 18 ("Robertson, James, 1950–53"), Anna Freud Papers.

23. Robertson to Anna Freud, November 21, 1954, box 86, folder 19 ("Robertson, James 1954–56"), Anna Freud Papers.

24. Helen Leland Witmer and Ruth Kotinsky, eds., *Personality in the Making: The Fact-Finding Report of the Midcentury White House Conference on Children and Youth* (New York: Harper and Row, 1952), xvi.

25. Ibid., 4–5; Erik Erikson, *Childhood and Society* (New York: Norton, 1950).

26. Witmer and Kotinsky, *Personality in the Making*, 93–96.

27. Pitirim A. Sorokin and Robert C. Hanson, "The Power of Creative Love," in *The Meaning of Love*, ed. Ashley Montagu (New York: Julian Press, 1953), 125; Ashley Montagu, "The Origin and Meaning of Love," in Montagu, *Meaning of Love*, 4, 4, 5, 18, 19. On Montagu's views on human nature, see Nadine Weidman, "An Anthropologist on TV: Ashley Montagu and the Biological Basis of Human Nature, 1945–1960," in Mark Solovey and Hamilton Cravens, eds. *Cold War Social Science* (New York: Palgrave Macmillan, 2012), 215–32.

28. Myrdal and Klein, *Women's Two Roles*, 125, 126, 127.

29. Clipping in Bowlby's Papers, PP/BOW/K.11/28.

30. "That Woman in Gray Flannel: A Debate," *New York Times*, February 12, 1956; Sloan Wilson, *The Man in the Gray Flannel Suit* (New York: Simon and Schuster, 1955).

31. "Should Mothers of Young Children Work?" *Ladies' Home Journal* 75 (November 1958): 58–59, 154–56, 158–61.

32. Ashley Montagu, "The Triumph and Tragedy of the American Woman," *Saturday Review* 41 (September 27, 1958): 14, 35, 34, 14.

33. Michel, *Children's Interests*, 155.

34. Margolis, *Mothers and Such*, 219.

35. Bowlby, *Child Care*, 43–44.

36. Ibid., 50, 41, 43.

37. Abraham Myerson, "Let's Quit Blaming Mom," *Science Digest* 29 (1951): 11; William H. Sewell, "Infant Training and the Personality of the Child," *American Journal of Sociology* 58 (1952): 150.

38. Harold Orlansky, "Infant Care and Personality," *Psychological Bulletin* 46 (1949): 12, used in Samuel R. Pinneau, "A Critique of the Articles by Margaret Ribble," *Child Development* 21 (1950): 222.

39. L. Joseph Stone, "A Critique of Studies of Infant Isolation," *Child Development* 25 (1954): 14.

40. Samuel R. Pinneau, "The Infantile Disorders of Hospitalism and Anaclitic Depression," *Psychological Bulletin* 52 (1955): 429–52.

41. Ibid., 447; see also ibid., 453ff. for Spitz's reply.

42. John Bowlby et al., "The Effects of Mother-Child Separation: A Follow-up Study," *British Journal of Medical Psychology* 29 (1956): 242.

43. John Bowlby, "A Note on Mother-Child Separation as a Mental Health Hazard," *British Journal of Medical Psychology* 31 (1958): 248.

44. Bowlby, *Child Care*, 69; E. F. M. Durbin and John Bowlby, *Personal Aggressiveness and War* (New York: Columbia University Press, 1939), includes an appendix on the social life of monkeys and apes.

45. Sigmund Freud, "Instincts and Their Vicissitudes" (1915); in *The Standard Edition of the Complete Psychological Works of Sigmund Freud*, translated under the general editorship of James Strachey in collaboration with Anna Freud and assisted by Alix Strachey and Alan Tyson, 24 vols. (London: Hogarth, 1953–74), 14:121–22. On this, see chapter 5.

46. John Bowlby, "An Ethological Approach to Research in Child Development," *British Journal of Medical Psychology* 30 (1957): 239.

47. John Bowlby, "The Nature of the Child's Tie to His Mother," *International Journal of Psychoanalysis* 39 (1958), 354.

48. Ibid., 358.

49. Ibid., 351.

50. Ibid., 362.

51. Ibid., 361.

52. Ibid., 367.

53. Ibid., 369.

54. Evelyn Fox Keller, "Physics and the Emergence of Molecular Biology: A History of Cognitive and Political Synergy," *Journal of the History of Biology* 23 (1990): 390. On the history and significance of objectivity in science see Daston and Galison, *Objectivity*.

55. Bowlby to Dugmore, October 19, 1957, PP/BOW/B.2/2. Bowlby Papers.

56. Bowlby sent reports of his trips for circulation to his research group. See: Copy of letter from Dr. Bowlby, Bowlby to Jock, January 18, 1958; and Bowlby to

Jock, April 11/13, 1958, which includes the passage quoted here. PP/BOW/B.2/2. Bowlby Papers.

57. Bowlby, "Nature of the Child's Tie," 369.

58. Ibid., 369–70.

59. Ibid., 370; emphasis added.

60. Ibid.

61. Bowlby, *Child Care*, 17.

62. Robert L. Griswold, *Fatherhood in America: A History* (New York: Basic Books, 1993), 186.

Chapter Four

1. Richard W. Burkhardt, *Patterns of Behavior: Konrad Lorenz, Niko Tinbergen, and the Founding of Ethology* (Chicago: University of Chicago Press, 2005); Paul E. Griffiths, "Instinct in the '50s: The British Reception of Konrad Lorenz's Theory of Instinctive Behavior," *Biology and Philosophy* 19 (2004): 609–31; Marga Vicedo, "Mother Love and Human Nature: A History of the Maternal Instinct" (PhD diss., Harvard University, 2005).

2. Niko Tinbergen, "Watching and Wondering," in *Studying Animal Behavior: Autobiographies of the Founders,* ed. Donald A. Dewsbury (Chicago: University of Chicago Press, 1989), 447.

3. Daniel S. Lehrman, "Comparative Behavior Studies," *Bird-Banding* 12 (1941): 86–87.

4. On Lehrman, see Jay S. Rosenblatt, "Daniel Sanford Lehrman, June 1, 1919–August 27, 1972," *Biographical Memoirs of the National Academy of Sciences* 66 (1995): 227–45; Colin Beer, "Was Professor Lehrman an Ethologist?" *Animal Behaviour* 23 (1975): 957–64. On animal studies at the American Museum of Natural History, see Gregg Mitman and Richard W. Burkhardt, "Struggling for Identity: The Study of Behavior in America, 1930–1945," in *The Expansion of American Biology*, ed. Keith R. Benson, Jane Maienschein, and Ronald Rainger (New Brunswick, NJ: Rutgers University Press, 1991), 164–94.

5. T. C. Schneirla and L. M. Chace, "Carpenter Ants," *Natural History* 60 (May 1951): 227–33; T. C. Schneirla and G. Piel, "The Army Ant," *Scientific American* 178 (June 1948): 16–23. On Schneirla, see Ethel Tobach and Lester R. Aronson, "T. C. Schneirla: A Biographical Note," in *Development and Evolution of Behavior: Essays in Memory of T. C. Schneirla*, ed. Lester R. Aronson, Ethel Tobach, Daniel S. Lehrman, and Jay S. Rosenblatt (San Francisco: Freeman, 1970), xi–xviii, and Gerard Piel, "The Comparative Psychology of T. C. Schneirla," in *Development and Evolution of Behavior*, ed. Aronson et al., 1–13. See also Howard R. Topoff's introduction to T. C. Schneirla, *Army Ants: A Study in Social Organization* (San Francisco: W. H. Freeman, 1971).

6. See T. C. Schneirla, "The Relationship between Observation and Experimentation in the Field Study of Behavior," *Annals of the New York Academy of Sciences* 51 (1950): 1022–44, and Schneirla, "A Consideration of Some Conceptual Trends in Comparative Psychology," *Psychological Bulletin* 49 (1952): 559–97.

7. Daniel Lehrman, "A Critique of Konrad Lorenz's Theory of Instinctive Behavior," *Quarterly Review of Biology* 28 (1953): 337–63.

8. Ibid., 341.

9. Ibid., 345.

10. Ibid., 343.

11. Ibid., 353.

12. Ibid., 354, criticizing Niko Tinbergen, "An Objectivist Study of the Innate Behaviour of Animals," *Bibliotheca Biotheoretica* 1 (1942): 39–98, and K. Lorenz, "Durch Domestikation verursachte Störungen arteigenen Verhaltens," *Zeitschrift für angewandte Psychologie und Charakterkunde* 59, nos. 1–2 (1940): 2–81. On Lorenz's Nazi ideology, see references in chapter 2.

13. Burkhardt, *Patterns of Behavior*, 385. Tinbergen as reported in Lorenz's autobiography for the Nobel Prize: http://www.nobelprize.org/nobel_prizes/medicine /laureates/1973/lorenz-autobio.html.

14. K. Lorenz, "The Objectivistic Theory of Instinct," in *L'instinct dans le comportement des animaux et de l'homme*, ed. M. Autuori et al. (Paris: Masson, 1956), 51, 56.

15. See Marga Vicedo, "Lehrman versus Lorenz: Facts, Semantics, and Politics in the Study of Animal Behavior," unpublished manuscript, 2012.

16. Daniel Lehrman, "On the Organization of Maternal Behavior and the Problem of Instinct," in *L'instinct*, ed. Autuori et al., 475–520. See also Lehrman, "The Physiological Basis of Parental Feeding Behavior in the Ring Dove (*Streptopelia risoria*)," *Behaviour* 7 (1955): 241–86; Daniel Lehrman and Rochelle P. Wortis, "Previous Breeding Experience and Hormone-Induced Incubation Behavior in the Ring Dove," *Science* 132 (1960): 1667–68; Lehrman, "Hormonal Regulation of Parental Behavior in Birds and Infrahuman Mammals," in *Sex and Internal Secretions*, ed. W. C. Young (Baltimore: Williams and Wilkins, 1961), 1268–82; Lehrman, "Interaction between Internal and External Environments in the Regulation of the Reproductive Cycle of the Ring Dove," in *Sex and Behavior*, ed. Frank A. Beach (New York: John Wiley, 1965), 355–80.

17. Lehrman, "On the Organization of Maternal Behavior," 483.

18. Lehrman to Hebb, August 25, 1949; Hebb to Beach, September 22, 1949; Beach to Hebb, September 26, 1949; Hebb to Lehrman, September 30, 1949: "Please remember that I find your main criticism of Lorenz very effective, even if most of the letter above is devoted to disagreeing with other parts of your paper. I must say also, I am shocked to find that Lorenz is one of the racially minded— I did not know this about him—and you are naturally bound to give him hell on that." All in "Miscellaneous Correspondence Files ca 1953–1977," container 0001, file L, MG1045, Donald O. Hebb Papers, McGill University Archives, Montreal.

19. Donald O. Hebb, "Heredity and Environment in Mammalian Behaviour," *British Journal of Animal Behaviour* 1 (1953): 43, 45, 46–47.

20. Konrad Lorenz to Bill Thorpe, March 11, 1955, cited by Burkhardt, *Patterns of Behavior*, 402.

21. Frank A. Beach, "The Descent of Instinct," *Psychological Review* 62 (1955): 401–10.

22. Ibid., 404, 405.

23. Ibid., 406, 407, 409.

24. Ibid., 405.

25. Frank A. Beach, "Retrospect and Prospect," in *Sex and Behavior*, ed. Frank A. Beach (New York: John Wiley, 1965), 547. The articles Beach reviewed are E. S. Valenstein, W. Riss, and W. C. Young, "Experiential and Genetic Factors in the Organization of Sexual Behavior in Male Guinea Pigs," *Journal of Comparative and Physiological Psychology* 48 (1955): 397–403; and A. A. Gerall, "The Effect of Prenatal and Postnatal Injections of Testosterone on Prepubertal Male Guinea Pig Sexual Behavior," *Journal of Comparative and Physiological Psychology* 68 (1963): 92–95.

26. Beach, "Retrospect and Prospect," 557.

27. Ibid., 547.

28. Niko Tinbergen, "Appendix II: Autobiographical Sketches of Participants," in *Group Processes: Transactions of the First Conference, September 26–30, 1954, Ithaca, New York*, ed. Bertram Schaffner (New York: Josiah Macy Jr. Foundation, 1955), 311–12.

29. Niko Tinbergen, "Psychology and Ethology as Supplementary Parts of a Science of Behavior," in *Group Processes*, ed. Schaffner, 102, 111.

30. Niko Tinbergen, "On Aims and Methods of Ethology," *Zeitschrift für Tierpsychologie* 20 (1963): 414, 424, 425.

31. Ibid., 425, 426.

32. Robert Hinde, "Ethological Models and the Concept of 'Drive,'" *British Journal for the Philosophy of Science* 6 (1956): 321–22.

33. Hinde, "Ethological Models," 324, 325. See also Hinde, "Unitary Drives," *Animal Behaviour* 7 (1959): 130–40, and Hinde, "Energy Models of Motivation" (1960), in *Foundations of Animal Behavior: Classic Papers with Commentaries*, ed. Lynne D. Houck and Lee C. Drickamer (Chicago: University of Chicago Press, 1996), 513–27.

34. Hinde, "Ethological Models," 326.

35. Ibid., 330.

36. Hinde, "Unitary Drives," 75.

37. Lorenz to Eckhard Hess, July 8, 1970, Konrad Lorenz Papers, Konrad Lorenz Institute for Evolution and Cognition Research, Altenberg, Austria.

38. T. C. Schneirla and Jay Rosenblatt, "'Critical Periods' in the Development of Behavior," *Science* 139 (1963): 1112.

39. Julian Jaynes, "Imprinting: The Interaction of Learned and Innate Behavior:

I. Development and Generalization," *Journal of Comparative and Physiological Psychology* 49 (1956): 201–6; Jaynes, "Imprinting: The Interaction of Learned and Innate Behavior: II. The Critical Period," *Journal of Comparative and Physiological Psychology* 50 (1957): 6–10; Jaynes, "Imprinting: The Interaction of Learned and Innate Behavior: III. Practice Effects on Performance, Retention, and Fear," *Journal of Comparative and Physiological Psychology* 51 (1958): 234–37; Jaynes, "Imprinting: The Interaction of Learned and Innate Behavior: IV. Generalization and Emergent Discrimination," *Journal of Comparative and Physiological Psychology* 51 (1958): 238–42.

40. R. Hinde, "The Modifiability of Instinctive Behaviour," *Advancement of Science* 12 (1955): 21.

41. W. H. Thorpe, *Learning and Instinct in Animals* (Cambridge, MA: Harvard University Press, 1956), 358. See Howard S. Hoffman and Alan M. Ratner, "A Reinforcement Model of Imprinting: Implications for Socialization in Monkeys and Men," *Psychological Review* 80 (1973): 527–44, who argue that the phenomenon can be explained by behavioral principles.

42. Howard Moltz, "Imprinting: Empirical Basis and Theoretical Significance," *Psychological Bulletin* 57 (1960): 300.

43. For a review of those studies, see Moltz, "Imprinting."

44. Eckhard H. Hess, "Imprinting," *Science* 130 (1959): 133–41.

45. Robert Hinde, "The Nature of Imprinting," in *Determinants of Infant Behaviour*, vol. 2, ed. B. M. Foss (London: Methuen, 1963), 227–33; Robert Hinde, "Some Aspects of the Imprinting Problem," *Symposia of the Zoological Society of London* 8 (1962): 129–38; P. P. G. Bateson, "The Characteristics and Context of Imprinting," *Biological Reviews* 41 (1966), 177–217.

46. Konrad Lorenz, *Evolution and Modification of Behavior* (Chicago: University of Chicago Press, 1965), 2.

47. Ibid., 4.

48. Ibid., 5.

49. Ibid., 20–21.

50. Ibid., 31, 34–35.

51. Ibid., 79, 42.

52. Lehrman, "Ethology and Psychology," *Recent Advances in Biological Psychiatry* 4 (1963): 92.

53. See Rosenblatt, "Lehrman." For Lehrman's work, see Daniel Lehrman, "On the Origin of the Reproductive Cycle in Doves," *Transactions of the New York Academy of Sciences* 21 (1959): 682–88; Lehrman, "Hormonal Regulation."

54. Daniel Lehrman, "Semantic and Conceptual Issues in the Nature-Nurture Problem," in *Development and Evolution of Behavior*, ed. Aronson et al., 21.

55. Ibid., 28.

56. Ibid.

57. Ibid., 34.

58. Ibid.

59. Ibid.

60. Tinbergen to L. Carmichael, October 23, 1970, Leonard Carmichael Papers, 1898–1973, Mss. B. C212, Box 1, American Philosophical Society, Philadelphia.

61. P. P. G. Bateson and Peter H. Klopfer, "Preface," in *Whither Ethology?* vol. 8 of *Perspectives in Ethology*, ed. P. P. G. Bateson and Peter H. Klopfer (New York: Plenum Press, 1989), v–vi; Beer, "Was Professor Lehrman an Ethologist?" 961.

Chapter Five

1. John Bowlby, "The Nature of the Child's Tie to His Mother," *International Journal of Psychoanalysis* 39 (1958): 361; see chapter 3 above.

2. Robert Karen, *Becoming Attached: First Relationships and How They Shape Our Capacity to Love* (New York: Oxford University Press, 1998), 113. For the view that Bowlby had not been notified of this assault beforehand, see also Inge Bretherton, "The Origins of Attachment Theory: John Bowlby and Mary Ainsworth," *Developmental Psychology* 28 (1992): 759–75. For Bowlby's view on the reception of his work by psychoanalysts, see Bowlby to Yven Gauthier, March 14, 1984, PP/BOW/B.2/4, John Bowlby Papers, Western Manuscripts and Archives, Wellcome Library, London. Frank C. P. Van der Horst and René Van der Veer, "Separation and Divergence: The Untold Story of James Robertson's and John Bowlby's Theoretical Dispute on Mother-Child Separation," *Journal of the History of the Behavioral Sciences* 45 (2009): 241, also presents Karen's version of the surprise attack and speculates about who orchestrated it.

3. On Freud, see Peter Gay, *Freud: A Life for Our Time* (New York: Anchor Books, 1988); Frank Sulloway, *Freud: Biologist of the Mind* (Cambridge, MA: Harvard University Press, 1992). On the history of psychoanalysis, see Nathan G. Hale Jr., *The Rise and Crisis of Psychoanalysis in the United States: Freud and the Americans, 1917–1985* (New York: Oxford University Press, 1995); and Joseph Schwartz, *Cassandra's Daughter: A History of Psychoanalysis* (New York: Viking, 1999).

4. See Sigmund Freud, *An Outline of Psychoanalysis* (1938), in *The Standard Edition of the Complete Psychological Works of Sigmund Freud*, trans. under the general editorship of James Strachey in collaboration with Anna Freud and assisted by Alix Strachey and Alan Tyson, 24 vols. (London: Hogarth, 1953–74), 23:144–47 (hereafter cited as *SE*).

5. Sigmund Freud, "Instincts and Their Vicissitudes" (1915), in *SE*, 14:118ff.

6. Freud, *Outline of Psychoanalysis*, in *SE*, 23:144–47. Freud had first elaborated these views in the early 1920s.

7. Freud, "Instincts and Their Vicissitudes," 122–23.

8. Ibid., 124.

9. Freud, *Outline of Psychoanalysis*, 148.

10. Freud, "Instincts and Their Vicissitudes," 123.

11. James Strachey presents an overview of Freud's use of the word instinct

in an "Editor's Note" before his translation of *Triebe und Triebschicksale*, in *SE*, 14:111–16.

12. Sigmund Freud, "Some Psychical Consequences of the Anatomical Distinction between the Sexes" (1925) in *SE*, 19:256. See also Sigmund Freud, "Femininity" (1933) in *SE*, 22:128.

13. Freud, "Femininity," 133.

14. On Freud's views on a woman's psyche as well as the responses of supporters and critics to those views, see Mary Jo Buhle, *Feminism and Its Discontents: A Century of Struggle with Psychoanalysis* (Cambridge, MA: Harvard University Press, 1998). See especially chap. 2 on Freud's views on femininity and discussion of Freud's "biologism." See also Lisa Appignanesi and John Forrester, *Freud's Women* (London: Phoenix, 2005)

15. Karen Horney, "Premenstrual Tension," in *Feminine Psychology* (1931; reprint, New York: Norton, 1967), 106.

16. Sigmund Freud to Dr. Carl Müller-Braunschweig, July 21, 1935. In Donald L. Burnham, "Freud and Female Sexuality: A Previously Unpublished Letter," *Psychiatry* 34 (August 1971): 328–29.

17. Karen Horney, "Premenstrual Tension;" Helene Deutsch, *Motherhood*, vol. 2 of *The Psychology of Women: A Psychoanalytic Interpretation* (New York: Grune and Stratton, 1945); Alice Balint, "Love for the Mother and Mother Love," *International Journal of Psychoanalysis* 30 (1949): 251–59; see also Appignanesi and Forrester, *Freud's Women*.

18. Sulloway, *Freud*, 5. On the influence of biological thought on Freud, see also Lucille B. Ritvo, *Darwin's Influence on Freud: A Tale of Two Sciences* (New Haven, CT: Yale University Press, 1972).

19. Lorenz to Erwin Stresemann, April 19, 1938, cited in Richard W. Burkhardt Jr., *Patterns of Behavior: Konrad Lorenz, Niko Tinbergen, and the Founding of Ethology* (Chicago: University of Chicago Press, 2005), 240.

20. Karl Bühler, *Die Krise der Psychologie* (Jena: Gustav Fischer, 1927). Similarities between ethology and psychoanalysis were pointed out by J. D. Carthy, "Instinct," *New Biologist* 10 (1951): 95–105; J. S. Kennedy, "Is Modern Ethology Objective?" *British Journal of Animal Behaviour* 2 (1954): 12–19; and Ronald Fletcher, *Instinct in Man* (New York: International Universities Press, 1957; reprint, New York: Schocken Books, 1966).

21. Lorenz to William Thorpe, March 12, 1958. W.TH/M/201.307 Add. 8784 M 16, William Homan Thorpe Papers, Department of Manuscripts and University Archives, University Library, Cambridge, by permission of the Syndics of Cambridge University Library; see also the introduction to Konrad Lorenz, *King Solomon's Ring: New Light on Animal Ways* (New York: Crowell, 1952).

22. Reviewed in Mortimer Ostow, "Psychoanalysis and Ethology," *Journal of the American Psychoanalytic Association* 8 (1960): 526–34. Other discussions included Therese Benedek, "On the Organization of Psychic Energy: Instincts,

Drives and Affects," in *Mid-century Psychiatry: An Overview*, ed. Roy R. Grinker (Springfield, IL: Charles C. Thomas, 1953), 60–75; M. Autuori, ed., *L'instinct dans le comportement des animaux et de l'homme* (Paris: Masson, 1956); Max Schur, "Discussion: A Psychoanalyst's Comments," *American Journal of Orthopsychiatry* 31 (1961): 276–91; and Heinz Hartmann, "Comments on the Psychoanalytic Theory of Instinctual Drives," in *Essays in Ego Psychology* (London: Hogarth Press and Institute of Psychoanalysis, 1964), 69–89, as well as other of his writings. For a review of discussions a few years later, see Leonard S. Zegans, "An Appraisal of Ethological Contributions to Psychiatric Theory and Research," *American Journal of Psychiatry* 124 (1967): 729–39.

23. Bowlby to Yven Gauthier, March 14, 1984, PP/BOW/B.2/4, John Bowlby Papers.

24. Bowlby to Anna Freud, January 4, 1950 (the date on the letter mistakenly says 1949), box 8, folder 7 ("Bowlby, John, 1946–80"), Anna Freud Papers, Sigmund Freud Collection, Manuscript Division, Library of Congress, Washington, DC.

25. Bowlby to Anna Freud, December 9, 1949, ibid. On April 9, 1951, Bowlby to Anna Freud: "indefinitely postponed."

26. Bowlby to Anna Freud, undated, ibid. The dates for the talks were June 22 and 27, 1950.

27. Bowlby to Anna Freud, June 26, 1950, ibid.

28. Bowlby to Anna Freud, April 9, 1951, ibid.

29. James Robertson to Anna Freud, January 7, 1952 (the date on the letter mistakenly says 1951), box 86, folder 18 ("Robertson, James, 1950–53"), Anna Freud Papers.

30. Anna Freud to Robertson, February 11, 1953, ibid.

31. James Robertson, *A Two-Year-Old Goes to Hospital* (London: Tavistock Child Development Research Unit, 1952), filmstrip.

32. John Bowlby and James Robertson, "A Two-Year-Old Goes to Hospital: A Film," *Proceedings of the Royal Society of Medicine* 46, no. 6 (1953): 425–27. See also John Bowlby, James Robertson, and D. Rosenbluth, "A Two-Year-Old Goes to Hospital," *Psychoanalytic Study of the Child* 7 (1952): 82–94; James Robertson and John Bowlby, "Responses of Young Children to Separation from Their Mothers: II. Observations of the Sequences of Response of Children Aged 18 to 24 Months during the Course of Separation," *Courrier du Centre International de l'Enfance* 3 (1952): 131–42. Bowlby notes the change of the third phase from "denial" to "detachment" in note 4 of Bowlby, "Separation Anxiety," *International Journal of Psychoanalysis* 41 (1960): 90.

33. Anna Freud, "The Concept of the Rejecting Mother" (1954), in *The Writings of Anna Freud*, vol. 4, *Indications for Child Analysis and Other Papers, 1945–1956* (New York: International Universities Press, 1968), 590, 591, 602, 593.

34. Robertson to Anna Freud, June 11, 1955, box 86, folder 19 ("Robertson, James, 1954–56"), Anna Freud Papers.

35. Donald Winnicott to Bowlby, May 11, 1954, in F. Robert Rodman, *The Spontaneous Gesture: Selected Letters of D. W. Winnicott* (Cambridge, MA: Harvard University Press, 1987), 65, 66.

36. Bowlby to Anna Freud, October 27, 1958, box 8, folder 7 ("Bowlby, John, 1946–80"), Anna Freud Papers.

37. Anna Freud to Bowlby, October 30, 1958; Anna Freud to Bowlby, February 20, 1959, ibid.

38. Bowlby to Ruth S. Eissler (copy sent to Anna Freud), June 23, 1959, ibid. As Anna Freud reported to Bowlby, she wrote to Eissler in November of that year supporting the publication of his paper and the plan to have it "with some answers" (Anna Freud to Eissler, November 2, 1959, ibid.).

39. John Bowlby, "Grief and Mourning in Infancy and Early Childhood," *Psychoanalytic Study of the Child* 15 (1960): 9–52.

40. Karen, *Becoming Attached*, 113. See also Bretherton, "Origins of Attachment Theory," and Van der Horst and Van der Veer, "Separation and Divergence," 241.

41. Bowlby to Anna Freud, December 11, 1959, box 8, folder 7 ("Bowlby, John, 1946–80"), Anna Freud Papers.

42. Bowlby to Anna Freud, February 3, 1960; Bowlby to Anna Freud, April 26, 1960, ibid.

43. Bowlby used "denial" as the third stage in "early drafts" of his paper "Separation Anxiety," but he changed the name to "detachment" in the final version, as he explains in a note. In the text Bowlby notes that "detachment" is "more descriptive" and is also preferable to "withdrawal" because of some of the associations of the latter term. See Bowlby, "Separation Anxiety," 90–91.

44. Anna Freud, "Discussion of John Bowlby's Work on Separation, Grief, and Mourning," in *The Writings of Anna Freud*, vol. 5, *Research at the Hampstead Child-Therapy Clinic and Other Papers, 1956–1965* (New York: International Universities Press, 1969), 167–86. A note states: "The remarks contained in Part I are here published for the first time. They were presented before the British Psychoanalytical Society, on November 5, 1958, when John Bowlby read his paper on 'Separation Anxiety.'"

45. Freud, "Discussion of John Bowlby's Work," 168. It is not clear why Anna Freud gave eighty as the number of children: the reports included a higher number. On Anna Freud's use of observations in her work, see Nick Midgley, "Anna Freud: the Hampstead War Nurseries and the Role of the Direct Observations of Children in Psychoanalysis," *International Journal of Psychoanalysis* 88 (2007): 939–59. On the types and roles of observations in science, see Lorraine Daston and Elizabeth Lunbeck, eds., *Histories of Scientific Observation* (Chicago: University of Chicago Press, 2011).

46. Freud, "Discussion of John Bowlby's Work," 169, 169–70.

47. Ibid., 170–71.

48. Anna Freud, "Discussion of Dr. John Bowlby's Paper," *Psychoanalytic Study of the Child* 15 (1960): 53.

49. Ibid., 53–54.

50. Ibid., 54.

51. Ibid., 56.

52. Ibid., 59.

53. Ibid.

54. Bowlby to Anna Freud, January 4, 1961, box 8, folder 7 ("Bowlby, John, 1946–80"), Anna Freud Papers.

55. Anna Freud and Dorothy Burlingham, *Infants without Families: Reports on the Hampstead Nurseries* (1943), in *The Writings of Anna Freud* (New York: International Universities Press, 1973), 3:12, 22; see also 40.

56. Ibid., 208.

57. Stephen M. Wittenberg and Lewis M. Cohen, "Max Schur, M.D., 1897–1969," *American Journal of Psychiatry* 159 (2002): 216.

58. Anna Freud to Schur, March 9, 1955, box 93, folder 17 ("Schur, Max, 1957–62"), Anna Freud Papers. Their correspondence is in German. All translations are mine.

59. Anna Freud to Schur, October 13, 1958; Schur to Anna Freud, February 14, 1959, ibid.

60. Schur to Anna Freud, January 14, 1960, ibid.

61. Schur to Anna Freud, April 6, 1960, ibid.

62. Max Schur, "Discussion of Dr. John Bowlby's Paper," *Psychoanalytic Study of the Child* 15 (1960): 63.

63. Ibid., 64, 64–65, 67; emphasis in the original.

64. Ibid., 67.

65. Anna Freud to Schur, May 10, 1960, box 93, folder 17 ("Schur, Max, 1957–62"), and Bowlby to Anna Freud, January 4, 1961, Box 8, folder 7 ("Bowlby, John, 1946–80"), both in Anna Freud Papers.

66. Bowlby to Anna Freud, December 30, 1960, box 8, folder 7 ("Bowlby, John, 1946–80"), Anna Freud Papers.

67. John Bowlby, "An Ethological Approach to Research in Child Development," *British Journal of Medical Psychology* 30 (1957), 234.

68. Schur to Anna Freud, January 27, 1961, box 93, folder 17 ("Schur, Max, 1957–62"), Anna Freud Papers.

69. (Copy of) Bowlby to Schur, December 30, 1960, box 8, folder 7 ("Bowlby, John, 1946–80"), Anna Freud Papers.

70. (Copy of) Eissler to Bowlby, May 10, 1961; (copy of) Bowlby to Eissler, May 26, 1961; Anna Freud to Bowlby, May 26, 1961; all in ibid.

71. Anna Freud to Schur, February 4, 1961, box 93, folder 17 ("Schur, Max, 1957–62"), Anna Freud Papers.

72. Bowlby, "Note on Dr. Max Schur's comments on Grief and Mourning in Infancy and Early Childhood," *Psychoanalytic* Study *of the Child* 16 (1961): 206–8; Schur to Anna Freud, May 8, 1961; Anna Freud to Schur, May 29, 1961, ibid.

73. Max Schur to Anna Freud, February 9, 1962, ibid.

74. René Spitz, *A Genetic Field Theory of Ego Formation: Its Implications for Pathology* (New York: International Universities Press, 1959), 30ff.

75. Spitz, "Introduction" (of Lorenz's talk), box M2116, folder 7 ("Lorenz 1970"), René Spitz Papers, Archives of the History of American Psychology, Center for the History of Psychology, University of Akron, Akron, OH.

76. Spitz, "Ethology" (final speech), dated 11/23/59, box M2115, folder 4 ("1959 PAPERS II"), Spitz Papers.

77. Spitz, "Discussion of Dr. Bowlby's Paper," *Psychoanalytic Study of the Child* 15 (1960): 87.

78. Ibid., 88.

79. Ibid., 91.

80. John Bowlby, "Psychoanalysis as Art and Science," *International Review of Psycho-Analysis* 6 (1979): 11. Bowlby noted that the data on children and mothers supported "a Winnicott type theory." See also Bowlby in Virginia Hunter, "John Bowlby: An Interview," *Psychoanalytic Review* 78 (1991): 170: "I always held the view that Winnicott and I were singing the same tune. We were essentially giving the same message, but again he didn't like my theoretical ideas. He and Fairbairn were always concerned with real life events. But, insofar as I used ideas derived from ethology, Winnicott didn't go along with that."

81. Karen, *Becoming Attached*, chap. 8.

Chapter Six

1. Harry Harlow, "The Nature of Love," *American Psychologist* 13 (1958): 673.

2. See, for example, Irenäus Eibl-Eibesfeldt, *Love and Hate: The Natural History of Behavior Patterns* (New York: Holt, Rinehart and Winston, 1972), chap. 10; Sarah Blaffer Hrdy, *Mother Nature: Maternal Instincts and the Shaping of the Species* (London: Vintage Books, 2000); Elaine Morgan, *The Descent of the Child: Human Evolution from a New Perspective* (London: Penguin, 1996), 117; Susan Allport, *A Natural History of Parenting: A Naturalist Looks at Parenting in the Animal World and Ours* (New York: Three Rivers Press, 1998), 167; Deborah Blum, *Love at Goon Park: Harry Harlow and the Science of Affection* (Cambridge, MA: Perseus, 2002); Robert Karen, *Becoming Attached: First Relationships and How They Shape Our Capacity to Love* (New York: Oxford University Press, 1998), chap. 9; Carl N. Degler, *In Search of Human Nature* (New York: Oxford University Press, 1991), 222; Matt Ridley, *Genome: The Autobiography of a Species in 23 Chapters* (New York: HarperCollins, 1999), 92.

3. Donna Haraway, *Primate Visions: Gender, Race, and Nature in the World of Modern Science* (New York: Routledge, 1989), 231, 240.

4. Letter of W. R. Miles, professor of experimental psychology at Stanford, to Lon H. Israel, Harlow's father, advising that his son "should change his name,"

May 28, 1930, Harry Harlow Papers, Harlow Primate Laboratory, University of Wisconsin–Madison, Madison, Wisconsin. Harlow's extensive archival materials are in the possession of his former longtime secretary and friend Helen A. LeRoy in Madison. Since the papers are not cataloged, I provide information on the folder and box only when available. I also have copies of all archival materials used here. On the problems of being Jewish in psychology at the time, see Andrew S. Winston, "'As His Name Indicates': R. S. Woodworth's Letters of Reference and Employment for Jewish Psychologists in the 1930s," *Journal of the History of the Behavioral Sciences* 32 (1996): 30–43.

5. Harry F. Harlow, "Mice, Monkeys, Men, and Motives," *Psychological Review* 60 (1953): 23.

6. On Harlow's life and work, see Blum, *Love at Goon Park*; H. A. LeRoy, "Harry Harlow: From the Other Side of the Desk," *Integrative Psychological and Behavioral Science* 42, no. 4 (2008): 348–53; H. A. LeRoy and G. A. Kimble, "Harry Frederick Harlow: And One Thing Led to Another . . . ," in *Portraits of Pioneers in Psychology*, ed. Gregory A. Kimble and Michael Wertheimer (Washington, DC: American Psychological Association, 2003), 5: 279–97; J. Sidowski and D. B. Lindsley, "Harry Frederick Harlow," *National Academy of Sciences Biographical Memoirs* 58 (1989): 219–57; S. Suomi and H. A. LeRoy, "In Memoriam: Harry F. Harlow (1905–1981)," *American Journal of Primatology* 2 (1982): 319–42.

7. For Harlow's own account of the development of his research on affects, see Harry F. Harlow, "Birth of the Surrogate Mother," in *Discovery Processes in Modern Biology*, ed. W. R. Klemm (Huntington, NY: Robert E. Krieger, 1977), 134–50.

8. For a description of the care practices in the lab, see A. J. Blomquist and Harry F. Harlow, "The Infant Rhesus Monkey Program at the University of Wisconsin Primate Laboratory," *Proceedings of the Animal Care Panel* 11 (1961): 57–64.

9. Harlow, "Nature of Love," 675.

10. Harry F. Harlow and R. Stagner, "Psychology of Feelings and Emotions: I. Theory of Feelings," *Psychological Review* 39 (1932): 570–89; Harlow and Stagner, "Psychology of Feelings and Emotions: II. Theory of Emotions," *Psychological Review* 40 (1933): 184–95.

11. Frank A. Beach to Harry Harlow, December 21, 1956; Harlow to Beach, December 28, 1956; Beach to Harlow, January 2, 1957; Harlow to Beach, January 5, 1957; Harlow to Ralph W. Tyler, August 22, 1957; all in Harlow Papers.

12. John Bowlby to Harry Harlow, August 8, 1957; Harlow to Bowlby, August 16, 1957. Bowlby referred to a draft of his paper "The Nature of the Child's Tie to His Mother," *International Journal of Psychoanalysis* 39 (1958): 350–73. On October 3, 1957, Harlow asked Bowlby for more copies. On Bowlby's visit, see Bowlby to Harlow, May 14, 1958; Harlow to Bowlby, May 23, 1958; Bowlby to Harlow, June 26, 1958; all in Harry Harlow Papers. On Bowlby and Harlow's correspondence, see Frank C. P. Van der Horst, Helen LeRoy, and René Van der Veer, "'When Strang-

ers Meet': John Bowlby and Harry Harlow on Attachment Behavior," *Integrative Psychological and Behavioral Science* 42 (2008): 370–88.

13. Harry F. Harlow and R. R. Zimmerman, "The Development of Affectional Responses in Infant Monkeys," *Proceedings of the American Philosophical Society* 102, no. 5 (1958): 501; Harlow and Zimmerman, "Affectional Responses in the Infant Monkey," *Science* 130 (1959): 421–32.

14. Harlow, "Nature of Love," 676, 674.

15. Ibid., 674.

16. Ibid., 676.

17. Ibid., 677. See also Harlow and Zimmerman, "Development of Affectional Responses"; Harlow and Zimmerman, "Affectional Responses"; Harry F. Harlow, "Basic Social Capacity of Primates," *Human Biology* 31 (1959): 40–53; Harlow, "Love in Infant Monkeys," *Scientific American* 200, no. 6 (1959), 68–74; Harlow, "Affectional Behavior in the Infant Monkey," in *The Central Nervous System and Behavior,* ed. Mary A. B. Brazier (New York: Josiah Macy Jr. Foundation, 1960), 307–57; and Harlow, "Of Love in Infants: Studies of Monkeys Show How Affection for Mothers Is Formed," *Natural History* 69, no. 5 (1960): 18–23.

18. Harlow, "Nature of Love," 678.

19. Ibid., 679–80.

20. Ibid., 684.

21. Ibid., 677–78.

22. Edward Tolman to Harlow, September 10, 1958, Harlow Papers.

23. Clippings in Harlow Papers. Three of Harlow's poems appeared in the *New York Times,* October 19, 1958, SM50.

24. Elaine Tyler May, *Homeward Bound: American Families in the Cold War Era* (New York: Basic Books, 1988); Nancy F. Cott, *Public Vows: A History of Marriage and the Nation* (Cambridge, MA: Harvard University. Press, 2000); Rebecca Jo Plant, *Mom: The Transformation of Motherhood in Modern America* (Chicago: University of Chicago Press, 2010); Steven Mintz and Susan Kellogg, *Domestic Revolutions: A Social History of American Family* Life (New York: Free Press, 1988); Jessica Weiss, *To Have and to Hold: Marriage, the Baby Boom, and Social Change* (Chicago: University of Chicago Press, 2000).

25. Harlow, "Nature of Love," 685.

26. Marston Bates, "Life with Mother" (Animal World), *New York Times,* October 19, 1958, SM44.

27. *New York Times,* clippings from Harlow Papers.

28. *Detroit Free Press,* May 1, 1959; clipping enclosed in letter from Mrs. D. L. Jones to Harry Harlow, May 2, 1959, in Harlow Papers.

29. Polly Parker to Harry Harlow, May 22, 1958, Harlow Papers.

30. Letter of Mrs. Albert Viall to the Office of the President, February 2, 1966, copy in Harlow Papers.

31. Maynard Kniskern, "'Love' in Baby Monkeys," *Springfield (OH) News-Sun,* September 7, 1958, 10a.

32. Harlow and Zimmerman, "Affectional Responses," 431.

33. Discussion of Harlow's "Affectional Behavior in the Infant Monkey," 344–45.

34. Solomon D. Kaplan, "Monkey Tricks and Voodoo Psychology," *American Psychologist* 15 (1960): 219.

35. Harry F. Harlow, "A Behavioral Approach to Psychoanalytic Theory," *Science and Psychoanalysis* 7 (1964): 100; see also Harry F. Harlow and Margaret K. Harlow, "Effects of Various Mother-Infant Relationships on Rhesus Monkey Behaviors," in *Determinants of Infant Behaviour: Proceedings of the Tavistock Seminar on Mother-Infant Interaction*, vol. 4, ed. Brian M. Foss (London: Methuen, 1969), 15–36.

36. Harlow, "Behavioral Approach," 107.

37. Harry F. Harlow, "Early Social Deprivation and Later Behavior in the Monkey," in *Unfinished Tasks in the Behavioral Sciences,* ed. Arnold Abrams, Harry H. Garner, and James E. P. Toman (Baltimore: Williams and Wilkins, 1964), 154; Harlow, "Monkeys, Men, Mice, Motives, and Sex," in *Psychological Research: The Inside Story*, ed. Michael H. Siegel and H. Philip Zeigler (New York: Harper and Row, 1976), 3.

38. Senator Harry F. Byrd as quoted in the *Minneapolis Morning Tribune*, February 3, 1962; cited in a letter from Donald W. Hastings to Senator Byrd, February 5, 1962. Undated clipping from *Arizona Election News* enclosed in letter from Israel Goldiamond to Harry Harlow, October 30, 1962. Both in Harlow Papers.

39. *Chicago Sunday Tribune*, February 4, 1962, clipping in Harlow Papers.

40. Harlow to Edwin B. Fred, President Emeritus, University of Wisconsin, February 6, 1968, Harlow Papers.

41. Letters to *Chicago Daily News*, February 7, 1962, clipping in Harlow Papers.

42. United States of America, *Congressional Record-House*, March 27, 1962, vol. 108, pt. 4, 5146 (Washington, DC: US Government Printing Office, 1962). Second quotation on p. 3 of the "Summary Statement concerning National Institute of Mental Health Support of Dr. Harry Harlow" (four pages, dated January 26, 1962, enclosed in a letter from Philip Sapir to Harry Harlow, February 9, 1962). Harlow thanked Melvin Laird for his support on April 4, 1962. Letters in Harlow Papers. See also Marjorie Hunter, "U.S. Aides Defend Monkey Research," *New York Times*, February 3, 1962, 36.

43. Harry F. Harlow, "The Development of Affectional Patterns in Infant Monkeys," in *Determinants of Infant Behaviour: Proceedings of the Tavistock Seminar on Mother-Infant Interaction,* vol. 1, ed. Brian. M. Foss (London: Methuen, 1961), 84.

44. Harlow, "Development of Affectional Patterns." See also Harry F. Harlow and Margaret K. Harlow, "A Study of Animal Affection," *Natural History* 70 (1961): 48–55.

45. Harlow, "Development of Affectional Patterns," 85.

46. Emma Harrison, "Monkey Mothers Upheld in Tests," *New York Times*, Sep-

tember 4, 1961, 17; Leonard Engel, "The Troubled Monkeys of Madison," *New York Times*, January 29, 1961, 62, 64; Emma Harrison, "Mother Trouble Affects Monkeys," *New York Times*, May 10, 1962, 75.

47. R. Spitz to Harlow, February 13, 1961; Harlow to Spitz, February 20, 1961; Harlow to Spitz, July 20, 1961; Harlow to Spitz, January 22, 1963, all in Harlow Papers.

48. Mary D. Ainsworth, "The Effects of Maternal Deprivation: A Review of Findings and Controversy in the Context of Research Strategy," in World Health Organization, *Deprivation of Maternal Care: A Reassessment of Its Effects* (Geneva: WHO, 1962), 128. Ainsworth cites Harlow, "Nature of Love"; H. F. Harlow, "Primary Affectional Patterns in Primates," *American Journal of Orthopsychiatry* 30 (1960): 676–84; and Harlow, "Development of Affectional Patterns."

49. Harry F. Harlow and Margaret K. Harlow, "The Effect of Rearing Conditions on Behavior," *Bulletin of the Menninger Clinic* 26 (1962): 213.

50. B. Seay, E. Hansen, and H. F. Harlow, "Mother-Infant Separation in Monkeys," *Journal of Child Psychology and Psychiatry and Allied Disciplines* 3 (1962): 123.

51. Seay, Hansen, and Harlow, "Mother-Infant Separation," 131, 130.

52. James H. Middlekauff to Harry Harlow, January 18, 1960 [*sic*; the correct date was 1961]; Harlow to Middlekauff, January 23, 1961, Harlow Papers.

53. Harlow to Middlekauff, January 23, 1961, Harlow Papers.

54. Harlow, "Primary Affectional Patterns," 676.

55. Harlow and Harlow, "Study of Animal Affection," 55.

56. Harry F. Harlow, "Affectional Systems of Monkeys, Involving Relations between Mothers and Young," *International Symposium on Comparative Medicine Proceedings* (New York: Eaton Laboratories, 1962), 6.

57. Ibid., 7, 8, 9.

58. Ibid., 9.

59. Ibid., 10.

60. Harry F. Harlow, "Motivation in Monkeys—and Men," in *Psychology and Life*, 6th ed., ed. Floyd Leon Ruch (Chicago: Scott, Foresman, 1963), 592, 591, 593.

61. Harry F. Harlow and B. Seay, "Affectional Systems in Rhesus Monkeys," *Journal of the Arkansas Medical Society* 61, no. 4 (1964): 110.

62. Harlow, "Early Social Deprivation," 168.

63. Ibid., 172–73; see also B. Seay, B. K. Alexander, and H. F. Harlow, "Maternal Behavior of Socially Deprived Rhesus Monkeys," *Journal of Abnormal and Social Psychology* 69 (1964): 345–54.

64. Stephen J. Suomi and Harry F. Harlow, "Abnormal Social Behavior in Young Monkeys," in *Exceptional Infant: Studies in Abnormalities,* vol. 2, ed. Jerome Hellmuth (New York: Brunner Mazel, 1971), 483–529; William T. McKinney Jr., Stephen J. Suomi, and Harry F. Harlow, "Studies in Depression," *Psychology Today* 4, no. 12 (1971): 61–63.

65. Harlow, "Behavioral Approach," 107.

66. "Child Adjustment Linked to Friends," *New York Times*, October 15, 1962, 45.

67. Harlow, "Early Social Deprivation," 171.

68. Harlow, "Motivation in Monkeys," 594.

69. Harry F. Harlow and Margaret K. Harlow, "The Affectional Systems," in *Behavior of Nonhuman Primates*, vol. 2, ed. H. F. Harlow, A. M. Schrier, and F. Stollnitz (New York: Academic Press, 1965), 287–334.

70. Harry F. Harlow and Margaret K. Harlow, "Developmental Aspects of Emotional Behavior," in *Psychological Correlates of Emotion*, ed. P. Black (New York: Academic Press, 1970), 56; see also Harlow and Harlow, "Affectional Systems"; and Harry F. Harlow, "The Primate Socialization Motives," *Transactions and Studies of the College of Physicians of Philadelphia*, ser. 4, 33 (1966): 224–37.

71. John Bowlby, "By Ethology Out of Psycho-analysis: An Experiment in Interbreeding," *Animal Behaviour* 28 (1980): 655.

72. Harry F. Harlow, Margaret K. Harlow, R. O. Dodsworth, and G. L. Arling, "Maternal Behavior of Rhesus Monkeys Deprived of Mothering and Peer Associations in Infancy," *Proceedings of the American Philosophical Society* 110, no. 1 (1966): 61.

73. Degler, *In Search of Human Nature*, 222; Ridley, *Genome*, 92. On Harlow's views on instincts, see Marga Vicedo, "The Evolution of Harry Harlow: From the Nature to the Nurture of Love," *History of Psychiatry* 21, no. 2 (2010), esp. 10–13.

74. Haraway, *Primate Visions*, 408, 241.

Chapter Seven

1. On Ainsworth, see Inge Bretherton and Mary Main, "Mary Dinsmore Salter Ainsworth (1913–1999)," *American Psychologist* 55 (2000): 1148–49. See also Robert Karen, *Becoming Attached: First Relationships and How They Shape Our Capacity to Love* (New York: Oxford University Press, 1998), who says Ainsworth progressed from "capable adherent" to "equal colleague" to "acknowledged partner"; Frank C. P. Van der Horst, *John Bowlby: From Psychoanalysis to Ethology* (Chichester, UK: Wiley-Blackwell, 2011), 152, presents Ainsworth as a "founder of the attachment paradigm" along with Bowlby; Susan van Dijken, *John Bowlby: His Early Life* (London: Free Association, 1998), 1, says Bowlby was "together with Mary Ainsworth, the founder of attachment theory"; and Jeremy Holmes, *John Bowlby and Attachment Theory* (London: Routledge, 1993), 26, also calls Ainsworth the "co-founder of Attachment Theory."

2. See Mary D. Salter Ainsworth, *Infancy in Uganda: Infant Care and the Growth of Love* (Baltimore: Johns Hopkins Press, 1967); John Bowlby, *Attachment*, vol. 1 of *Attachment and Loss* (New York: Basic Books, 1969).

3. Mary D. Salter, *An Evaluation of Adjustment Based upon the Concept of*

Security (Toronto: University of Toronto Press, 1940). In a 1969 paper Ainsworth noted that she got the concept of "secure base from which to explore" from Blatz. See Mary D. Salter Ainsworth and Barbara A. Wittig, "Attachment and Exploratory Behavior of One-Year-Olds in a Strange Situation," in *Determinants of Infant Behaviour*, vol. 4, ed. B. M. Foss (London: Methuen, 1969), 112. She repeated this statement in a letter to John Bowlby on February 23, 1971, noting that the concept was passed on via "oral tradition" and she could not find it either in his publications or in her own earlier writings. Ainsworth to Bowlby, February 23, 1971, box M3168, folder 3, "Correspondence-Bowlby (continued), 1968–1971," Mary Ainsworth Papers, Archives of the History of American Psychology, Center for the History of Psychology, University of Akron, Akron, OH. In *Patterns of Attachment*, Ainsworth mentioned having heard Blatz use the concept when she was a student in one of his courses in 1934–35. See Mary D. Salter Ainsworth, Mary C. Blehar, Everett Waters, and Sally Wall, *Patterns of Attachment: A Psychological Study of the Strange Situation* (Hillsdale, NJ: Erlbaum, 1978), ix.

4. Ainsworth, *Infancy in Uganda*, 3.

5. Ainsworth et al., *Patterns of Attachment*, viii.

6. For the plan for the joint book and other documents relevant to CS1, including Ainsworth's draft and minutes of the group meetings and the group's comments, see PP/BOW/D.3/2 "CS1: Synopses etc., 1953–55" and PP/BOW/D.3/3 "CS1: Team comment etc., 1954." See also Bowlby to Ainsworth, March 7, 1955; Ainsworth to Bowlby, March 10, 1955, PP/BOW/D.3/2 "CS1 Synopses etc., 1953–55." All in John Bowlby Papers, Wellcome Library, London.

7. Bowlby to Ainsworth, February 14, 1958; Ainsworth to Bowlby, February 21, 1958; Bowlby to Ainsworth, February 26, 1958. All in box M3168, folder 1 ("Correspondence Bowlby"), Ainsworth Papers. For a detailed analysis of this episode, see Marga Vicedo, "The Making of Attachment: The Theoretician, the Experimenter, and the Social Worker," unpublished manuscript, 2012.

8. Lawrence Casler, "Maternal Deprivation: A Critical Review of the Literature," *Society for Research in Child Development* 26 (1961): 9. On Pinneau, see chapter 3.

9. Casler, "Maternal Deprivation." Also see Lawrence Casler, "Perceptual Deprivation in Institutional Settings," in *Early Experience and Behavior*, ed. Grant Newton and Seymour Levine (Springfield, IL: Charles C. Thomas, 1968), 573–626; Neil O'Connor, "The Evidence for the Permanently Disturbing Effects of Mother-Child Separation," *Acta Psychologica* 12 (1956): 174–91.

10. World Health Organization, *Deprivation of Maternal Care: A Reassessment of Its Effects* (Geneva: World Health Organization, 1962).

11. Dane G. Prugh and Robert G. Harlow, "'Masked Deprivation' in Infants and Young Children," in World Health Organization, *Deprivation of Maternal Care*, 13; R. G. Andry, "Paternal and Maternal Roles and Delinquency," in World Health Organization, *Deprivation of Maternal Care*, 35; Serge Lebovici, "The Con-

cept of Maternal Deprivation: A Review of Research," in World Health Organization, *Deprivation of Maternal Care*, 90.

12. Barbara Wootton, "A Social Scientist's Approach to Maternal Deprivation," in World Health Organization, *Deprivation of Maternal Care*, 63, 68.

13. Margaret Mead, "A Cultural Anthropologist's Approach to Maternal Deprivation," in World Health Organization, *Deprivation of Maternal Care*, 55.

14. Mead, "Cultural Anthropologist's Approach," 57.

15. Ibid., 57, 58.

16. See Dr. E. E. Krapf to Ainsworth, September 20, 1961, and Ainsworth to Krapf, September 25, 1961, box M3167, folder 3, Ainsworth Papers. Krapf, who was chief of mental health at the WHO, also suggested changes, including asking her to address the criticisms Mead raised. Ainsworth's concern was to present "the Bowlby viewpoint accurately" rather than offering an independent critical evaluation of Bowlby's work.

17. Mary D. Ainsworth, "The Effects of Maternal Deprivation: A Review of Findings and Controversy in the Context of Research Strategy," in World Health Organization, *Deprivation of Maternal Care*, 105.

18. Ainsworth, "Effects of Maternal Deprivation," 146–47.

19. John Bowlby, *Child Care and the Growth of Love*, ed. Margery Fry (Harmondsworth, UK: Pelican, 1953), 75.

20. Ainsworth to Bowlby, June 22, 1965, box M3168, folder 2, "Correspondence (continued)—Bowlby 1964–67," Ainsworth Papers; original also in PP/BOW/D.3/8, Bowlby Papers.

21. Ainsworth to Bowlby, October 7, 1965, box M3168, folder 2, "Correspondence (continued)—Bowlby 1964–67," Ainsworth Papers.

22. For Robertson's emphasis on the limitations of his data, see "JR to Dr. Bowlby, CS1 chapter 2 (formerly chapter 3) 12.11.64," in PP/BOW/D.3/40: "CS1: Preface. Chapters 1, 2, 3. Comment 1965." Bowlby Papers. On scientific authorship, see Mario Biagioli and Peter Galison, eds. *Scientific Authorship: Credit and Intellectual Property in Science* (New York: Routledge, 2012).

23. Ainsworth to Bowlby, June 20, 1967, box M3168, folder 2, "1964–1967, Correspondence-Bowlby, continued," Ainsworth Papers.

24. Ainsworth to Bowlby, December 1, 1968, box M3170, folder 4, Ainsworth Papers.

25. Ainsworth to Bowlby, October 29, 1966, ibid.

26. Mary D. Salter Ainsworth and John Bowlby, "An Ethological Approach to Personality Development," *American Psychologist* 46 (1991): 333. See also Ainsworth, "Attachment as Related to Mother-Infant Interaction," *Advances in the Study of Behavior* 9 (1979): 2.

27. Mary D. Ainsworth, "The Development of Infant-Mother Interaction among the Ganda," in *Determinants of Infant Behaviour*, vol. 2, ed. B. M. Foss (London: Methuen, 1963), 67–104.

28. Ainsworth, *Infancy in Uganda*, 41.

29. Ibid., 38–39.

30. Ibid., 92, 94.

31. Ibid., 391, 392.

32. Ibid., 393.

33. Ibid.

34. Ibid., 397.

35. Ibid., 398.

36. Ibid., 449–51.

37. Ibid., 400, 451–52.

38. Ibid., 400.

39. Jean M. Arsenian, "Young Children in an Insecure Situation," *Journal of Abnormal and Social Psychology* 38 (1943): 225–49.

40. Ainsworth and Wittig, "Attachment and Exploratory Behavior," 136; Ainsworth to Bowlby, April 9. 1967, box M3168, folder 2 ("1964–1967, Correspondence-Bowlby, continued"), Ainsworth Papers; Ainsworth, "Attachment as Related to Mother-Infant Interaction"; Ainsworth, "Infant-Mother Attachment," *American Psychologist* 34 (1979): 932–37; Ainsworth et al., *Patterns of Attachment*.

41. Ainsworth to Bowlby, June 22, 1965, box M3168, folder 2: "1964–1967, Correspondence-Bowlby, continued," Ainsworth Papers.

42. Ainsworth, *Infancy in Uganda*. Bowlby's foreword (v) notes that her approach "is the natural history approach typical of the social anthropologist and of the ethologist." Bretherton and Main, "Mary Ainsworth," say: "While conducting naturalistic observations in the homes of Ganda women, Ainsworth suddenly realized the relevance of Bowlby's ethologically inspired ideas" (1148). However, Ainsworth herself recognized that "the full significance of what I observed and recorded in my field reports emerged only gradually, not merely in the process of analyzing my observations, but also in the course of reading, discussions with others interested in mother-infant interaction, and further research into the early development of attachment" (Ainsworth, *Infancy in Uganda*, ix).

43. Ibid., 399.

44. Ibid., 439.

45. Ibid., 429.

46. Ibid., 429–30.

47. Ibid., 436–37.

48. Ibid., 393.

49. Ainsworth and Wittig, "Attachment and Exploratory Behavior," 134.

50. Ainsworth et al., *Patterns of Attachment*, 150.

51. John Bowlby, *Attachment,* vol. 1 of *Attachment and Loss,* 2nd ed.(New York: Basic Books, 1982), 345–46, refers to Ainsworth and Wittig, "Attachment and Exploratory Behavior," and to Ainsworth et al., *Patterns of Attachment*.

52. Ainsworth, *Infancy in Uganda*, 387–88.

53. Ibid., 387; my emphasis.

54. Ainsworth, *Infancy in Uganda*, 437, 438.

55. Ibid., 432.

Chapter Eight

1. John Bowlby, *Attachment and Loss,* vol. 1, *Attachment* (New York: Basic Books, 1969), 345.

2. Copy of *Attachment* signed by Bowlby in May 1969, in Konrad Lorenz's library in Lorenz's house. Konrad Lorenz Institute for Evolution and Cognition Research, Altenberg, Austria.

3. Konrad Lorenz, *On Aggression* (New York: Harcourt Brace and World, 1966.) All quotations in this book are from the 1969 Bantam edition.

4. Ibid., 46.

5. Ibid., 238.

6. Ibid., 46.

7. Ibid., xiii–xiv.

8. Ashley Montagu, ed., *Man and Aggression* (New York: Oxford University Press, 1968).

9. Lorenz, *On Aggression*, x.

10. Spitz, untitled introduction to Lorenz's talk, January 11, 1970, box M2116, folder 7, "Lorenz 1970," René Spitz Papers, Archives of the History of American Psychology, Center for the History of Psychology, University of Akron, Akron, OH.

11. René A. Spitz, in collaboration with W. Godfrey Cobliner, *The First Year of Life: A Psychoanalytic Study of Normal and Deviant Development of Object Relations* (New York: International Universities Press, 1965), 206, 207, 209; emphasis in the original.

12. Ibid., 117, 118; emphasis in the original.

13. Konrad Lorenz, *Studies in Animal and Human Behavior,* vol. 1 (Cambridge, MA: Harvard University Press, 1970), xii; my emphasis.

14. Ibid.

15. Lorenz to E. Hess, July 13, 1964, Konrad Lorenz Papers, Konrad Lorenz Institute for Evolution and Cognition Research, Altenberg, Austria (my translation).

16. Lorenz to Wylie, April 17, 1969; Wylie to Lorenz, April 16, 1970, Philip Wylie Papers, Manuscripts Division, Department of Rare Books and Special Collections, Princeton University Library, Princeton, NJ. On Wylie, see references in chapter 1.

17. Philip Wylie, *The Magic Animal* (New York: Doubleday, 1968), 13, 90, 69. For a report of Lorenz's visit, see *Lerner Marine Laboratory Newsletter* 2, no. 2 (May 1970).

18. Konrad Lorenz, *Civilized Man's Eight Deadly Sins* (New York: Harcourt, 1973), 39. On eugenics, see Diane Paul, *Controlling Human Heredity: 1865 to the Present* (Atlantic Highlands, N.J: Humanities Press, 1995).

19. Joseph Alsop, "Profiles: A Condition of Enormous Improbability," *New Yorker*, March 8, 1969, 84–85, 45, 58.

20. Evelyn Fox Keller, *Refiguring Life: Metaphors of Twentieth-Century Biology* (New York: Columbia University Press, 1995), 19.

21. Alsop, "Profiles," 82.

22. "The Family, to Lorenz, Is All," *New York Times*, January 22, 1970.

23. "Rats, Apes, Naked Apes, Kipling, Instincts, Guilt, the Generations and Instant Copulation—a Talk with Konrad Lorenz," *New York Times*, July 5, 1970.

24. Richard W. Burkhardt, *Patterns of Behavior: Konrad Lorenz, Niko Tinbergen, and the Founding of Ethology* (Chicago: University of Chicago Press, 2005), chap. 10.

25. Eckhard Hess, "On Anthropomorphism," in *Unfinished Tasks in the Behavioral Sciences*, ed. Arnold Abrams, Harry H. Garner, and James E. P. Tolman (Baltimore: Williams and Wilkins, 1964), 174.

26. Lorenz quoted in Alsop, "Profiles," 64, 65.

27. Brendan Gill, "The Unfinished Man" (review of *Portnoy's Complaint*, by Philip Roth), *New Yorker*, March 8, 1969, 120.

28. Karolinska Institutet, "Physiology or Medicine 1973—Press Release," accessed September 11, 2012, http://nobelprize.org/nobel_prizes/medicine/laureates /1973/press.html.

29. Lorenz, *Civilized Man's Eight Deadly Sins*, 43, 48.

30. Spitz, *First Year of Life*, 300.

31. René Spitz, "Discussion of Dr. Lorenz's Paper," January 11, 1970. box M2116, folder 7, "Lorenz 1970," Spitz Papers.

32. Lorenz, *Civilized Man's Eight Deadly Sins*, 48, 53, 59.

33. Konrad Lorenz quoted in "Nobel Laureate Watches Fish for Clues to Human Violence," *New York Times*, May 8, 1977.

34. Bowlby, *Attachment*, xvii.

35. James Robertson and Joyce Robertson, "Young Children in Brief Separation: A Fresh Look," *Psychoanalytic Study of the Child* 26 (1971): 312–13.

36. John Bowlby, "By Ethology Out of Psycho-analysis: An Experiment in Interbreeding," *Animal Behaviour* 28 (1980): 655.

37. Bowlby, "By Ethology Out of Psycho-analysis," 650. Other scholars writing about attachment often note that it received support from Harlow. See, for example, Robert Karen, *Becoming Attached: First Relationships and How They Shape Our Capacity to Love* (New York: Oxford University Press, 1998), 122; on Harlow's experiments, see chapter 6.

38. Robert A. Hinde and Lynda Davies, "Removing Infant Rhesus from Mother for 13 Days Compared with Removing Mother from Infant," *Journal of Child Psychology and Psychiatry* 13 (1972): 227–37; Robert A. Hinde and L. McGinnis, "Some Factors Influencing the Effects of Temporary Mother-Infant Separation: Some Experiments with Rhesus Monkeys," *Psychological Medicine* 7 (1977): 197–212; Hinde, "Ethology and Attachment Theory," in *Attachment from Infancy*

to Adulthood: The Major Longitudinal Studies, ed. Klaus Grossmann, Karin Grossmann, and Everett Waters (New York: Guilford Press, 2005), 1–12.

39. Robert. A. Hinde, "Mother-Infant Separation and the Nature of Interindividual Relationships: Experiments with Rhesus Monkeys," *Proceedings of the Royal Society,* ser. B, 196 (1977): 49.

40. Bowlby, *Attachment*, 223.

41. Ibid., 39.

42. Robert Hinde to John Bowlby, June 29, 1965, in PP/BOW/K.4/5 "Attachment. Chapters 3–8: comments Child's Tie draft." Bowlby Papers.

43. Bowlby, *Attachment*, 51, 45.

44. Mary Ainsworth, "Object Relations, Dependency, and Attachment: A Theoretical Review of the Infant-Mother Relationship," *Child Development* 40 (1969): 971, 995.

45. Konrad Z. Lorenz, "The Fashionable Fallacy of Dispensing with Description," *Die Naturwissenschaften* 60 (1973): 7.

46. Ainsworth, "Object Relations, Dependency, and Attachment," 995–96.

47. Bowlby, *Attachment*, 54.

48. John Bowlby, *Child Care and the Growth of Love*, ed. Margery Fry (Harmondsworth, UK: Pelican, 1953), 181.

49. Bowlby in Evelyn S. Ringold, "Bringing Up Baby in Britain," *New York Times*, June 13, 1965.

50. Bowlby in Robert J. Trotter, "Human Behavior: Do Animals Have the Answer?" *Science News* 105 (1974): 279.

51. Bowlby, *Attachment*, 166.

52. Mary D. Salter Ainsworth, Mary C. Blehar, Everett Waters, and Sally Wall, *Patterns of Attachment: A Psychological Study of the Strange Situation* (Hillsdale, NJ: Lawrence Erlbaum, 1978), 9.

53. Mary D. Salter Ainsworth, "Infant-Mother Attachment," *American Psychologist* 34 (1979): 932. She claims (935) that it is a "mistaken interpretation" to say that Bowlby claimed an infant can become attached only to its mother. I addressed this issue earlier.

54. Mary D. Ainsworth, "Attachment as Related to Mother-Infant Interaction," *Advances in the Study of Behavior* 9 (1979): 37.

55. John Bowlby, *A Secure Base: Parent-Child Attachment and Healthy Human Development* (New York: Basic Books, 1988), 165.

56. Mary D. Salter Ainsworth and John Bowlby, "An Ethological Approach to Personality Development," *American Psychologist* 46 (1991): 333; my emphasis.

Conclusion

1. John Bowlby, *Attachment*, vol. 1 of *Attachment and Loss* (New York: Basic Books, 1969), 166.

2. John Bowlby, *Child Care and the Growth of Love*, ed. Margery Fry (Harmondsworth, UK: Pelican, 1953), 17.

3. For an insightful critical analysis of the biologizing of the affects, see Ruth Leys, "The Turn to Affect: A Critique," *Critical Inquiry* 37 (2011): 434–72.

4. For critical assessments of attachment theory and research, see Rochelle Paul Wortis, "The Acceptance of the Concept of the Maternal Role by Behavioral Scientists: Its Effects on Women," *American Journal of Orthopsychiatry* 41, no. 5 (1971): 733–46; D. W. Rajecki, Michael E. Lamb, and Pauline Obmascher, "Toward a General Theory of Infantile Attachment: A Comparative Review of Aspects of the Social Bond," *Behavioral and Brain Sciences* 3 (1978): 417–64; Michael E. Lamb et al., "Security of Infantile Attachment as Assessed in the 'Strange Situation': Its Study and Biological Interpretation," *Behavioral and Brain Sciences* 7 (1984): 127–47, and the accompanying "Open Peer Commentary," 147–71; Beverly Birns and Niza Ben-Ner, "Psychoanalysis Constructs Motherhood," in *The Different Faces of Motherhood*, ed. Beverley Birns and Dale F. Hay (New York: Plenum Press, 1988), 47–72; Jerome Kagan, *Unstable Ideas: Temperament, Cognition, and Self* (Cambridge, MA: Harvard University Press, 1989); John R. Morss, *The Biologising of Childhood: Developmental Psychology and the Darwinian Myth* (Hove, UK: Lawrence Erlbaum, 1990); Diane E. Eyer, *Mother-Infant Bonding: A Scientific Fiction* (New Haven, CT: Yale University Press, 1992); Eyer, *Motherguilt: How Our Culture Blames Mothers for What's Wrong with Society* (New York: Random House, 1996); Michael Rustin, "Attachment in Context," in *The Politics of Attachment: Towards a Secure Society*, ed. Sebastian Kraemer and Jane Roberts (London: Free Association Books, 1996), 212–28; and Michael Lewis, *Altering Fate: Why the Past Does Not Predict the Future* (New York: Guilford Press, 1997).

5. Eyer, *Motherguilt*, 88.

6. William Sears and Martha Sears, *The Baby Book*, 2nd ed. (1992; New York: Little, Brown, 2003), 411, 417; emphasis in the original. For a recent overview of Sears in the popular press, see Kate Pickert, "The Man Who Remade Motherhood," *Time*, May 21, 2012, 32–39. For other major defenders of attachment in child care, see the writings of pediatrician T. Berry Brazelton and psychologist Jay Belsky. On Brazelton, see Eyer, *Motherguilt*. For Belsky's views, see Jay Belsky and Laurence D. Steinberg, "The Effects of Day Care: A Critical Review," *Child Development* 49 (1978): 929–49. On Belsky, see Susan J. Douglas and Meredith W. Michaels, *The Mommy Myth: The Idealization of Motherhood and How It Has Undermined All Women* (New York: Free Press, 2004).

7. Ainsworth to Bowlby, April 9, 1967, box M3168, folder 2 ("Correspondence Bowlby"), Mary Ainsworth Papers, Archives of the History of American Psychology, Center for the History of Psychology, University of Akron, Akron, OH.

8. Robert Karen, *Becoming Attached: First Relationships and How They Shape Our Capacity to Love* (New York: Oxford University Press, 1998), 148.

9. Robert Kohler, *Lords of the Fly: Drosophila Genetics and the Experimental Life* (Chicago: The University of Chicago Press, 1994), 8.

10. Robert A. Hinde, "Attachment: Some Conceptual and Biological Issues," in *The Place of Attachment in Human Behavior*, ed. Colin Murray Parkes and Joan Stevenson-Hinde (New York: Basic Books, 1982), 70.

11. Ibid., 73. He also asks (65): "Does 'attachment' refer to an individual or a dyad?" And again (66), he wonders what the strange situation is measuring: categories of behavior, categories of infants, categories of relationships? Hinde criticized the extrapolations from suppositions about the environment of evolutionary adaptedness to the view that the behavior is normal and therefore desirable. See Robert A. Hinde and Joan Stevenson-Hinde, "Attachment: Biological, Cultural and Individual Desiderata," *Human Development* 33 (1990): 62–72, and Hinde, "Relationships, Attachment, and Culture: A Tribute to John Bowlby," *Infant Mental Health Journal* 12 (1991): 154–63. For an overview of his work, see Hinde, *Biological Bases of Human Social Behavior* (New York: McGraw-Hill, 1974).

12. Mary D. Salter Ainsworth and John Bowlby, "An Ethological Approach to Personality Development," *American Psychologist* 46 (1991): 336.

13. Elliott Sober, "Evolution, Population Thinking, and Essentialism," *Philosophy of Science* 47 (1980): 374.

14. Mary D. Salter Ainsworth, "Attachments and Other Affectional Bonds across the Life Cycle," in *Attachment across the Life Cycle,* ed. Colin Murray Parkes, Joan Stevenson-Hinde, and Peter Marris (New York: Routledge, 1991), 33.

15. John Byng-Hall, "An Appreciation of John Bowlby: His Significance for Family Therapy," *Journal of Family Therapy* 13 (1991): 6.

16. Daniel S. Lehrman, "Can Psychiatrists Use Ethology?" in *Ethology and Psychiatry*, ed. Norman F. White (Toronto: University of Toronto Press, 1974), 190.

17. Ibid., 190–91.

18. Ibid., 193. See I. Charles Kaufman, "Mother/Infant Relations in Monkeys and Humans: A Reply to Professor Hinde," in White, *Ethology and Psychiatry*, 47–68.

19. I. Charles Kaufman, "Biologic Considerations of Parenthood," in *Parenthood: Its Psychology and Psychopathology*, ed. E. James Anthony and Therese Benedek (1970; reprint, Boston: Little, Brown, 1996), 3–55.

20. Sarah Blaffer Hrdy, *Mother Nature: Maternal Instincts and the Shaping of the Species* (London: Vintage Books, 2000).

21. Hrdy, *Mother Nature*, 231. On using animals to think about human nature, see Lorraine Daston and Gregg Mitman, eds., *Thinking with Animals: New Perspectives on Anthropocentrism* (New York: Columbia University Press, 2005).

22. See Sarah Hall Sternglanz and Alison Nash, "Ethological Contributions to the Study of Human Motherhood," in Birns and Hay, *Different Faces of Motherhood*, 15–46.

Bibliography

Primary and Secondary Works

Ainsworth, Mary D. Salter. "The Effects of Maternal Deprivation: A Review of Findings and Controversy in the Context of Research Strategy." In World Health Organization, *Deprivation of Maternal Care: A Reassessment of Its Effects,* 97–165. Geneva: World Health Organization, 1962.

———. *Infancy in Uganda: Infant Care and the Growth of Love.* Baltimore: Johns Hopkins Press, 1967.

———. "Object Relations, Dependency, and Attachment: A Theoretical Review of the Infant-Mother Relationship." *Child Development* 40 (1969): 969–1025.

———. "Attachment as Related to Mother-Infant Interaction." *Advances in the Study of Behavior* 9 (1979): 1–51.

———. "Infant-Mother Attachment." *American Psychologist* 34 (1979): 932–37.

———. "Attachments and Other Affectional Bonds across the Life Cycle." In *Attachment across the Life Cycle*, edited by Colin Murray Parkes, Joan Stevenson-Hinde, and Peter Marris, 33–51. New York: Routledge, 1991.

Ainsworth, Mary D. Salter, Mary C. Blehar, Everett Waters, and Sally Wall. *Patterns of Attachment: A Psychological Study of the Strange Situation*. Hillsdale, NJ: Lawrence Erlbaum, 1978.

Ainsworth, Mary D. Salter, and John Bowlby. "An Ethological Approach to Personality Development." *American Psychologist* 46 (1991): 333–41.

Ainsworth, Mary D. Salter, and B. A. Wittig. "Attachment and Exploratory Behaviour of One-Year-Olds in a Strange Situation." In *Determinants of Infant Behaviour*, vol. 4, edited by Brian M. Foss, 111–36. London: Methuen, 1969.

Alexander, Franz. *Our Age of Unreason: A Study of the Irrational Forces in Social Life*. 1942. Rev. ed., Philadelphia: Lippincott, 1951.

Allport, Gordon W. "Scientific Models and Human Morals." *Psychological Review* 54 (1947): 182–92.

Allport, Susan. *A Natural History of Parenting: A Naturalist Looks at Parenting in the Animal World and Ours*. New York: Three Rivers Press, 1998.

Alsop, Joseph. "Profiles: A Condition of Enormous Improbability." *New Yorker*, March 8, 1969, 39–93.

Appignanesi, Lisa, and John Forrester. *Freud's Women*. London: Phoenix, 2005.

Apple, Rima D. *Perfect Motherhood: Science and Childrearing in America*. New Brunswick, NJ: Rutgers University Press, 2006.

Apple, Rima D., and Janet Golden, eds. *Mothers and Motherhood: Readings in American History*. Columbus: Ohio State University Press, 1997.

Andry, R. G. "Paternal and Maternal Roles and Delinquency." In World Health Organization, *Deprivation of Maternal Care: A Reassessment of Its Effects*, 31–44. Geneva: World Health Organization, 1962.

Arsenian, Jean M. "Young Children in an Insecure Situation." *Journal of Abnormal and Social Psychology* 38 (1943): 225–49.

Ash, Mitchell G. "Psychology and Politics in Interwar Vienna: The Vienna Psychological Institute, 1922–1942." In *Psychology in Twentieth-Century Thought and Society*, edited by M. G. Ash and W. R. Woodward, 143–64. Cambridge: Cambridge University Press, 1987.

Autuori, Mario, et al., eds. *L'instinct dans le comportement des animaux et de l'homme*. Paris: Masson, 1956.

Bach, William G. "The Influence of Psychoanalytic Thought on Benjamin Spock's *Baby and Child Care*." *Journal of the History of the Behavioral Sciences* 10 (1974): 91–94.

Badinter, Elisabeth. *Mother Love: Myth and Reality*. New York: Macmillan, 1981.

Balint, Alice. "Love for the Mother and Mother Love." *International Journal of Psychoanalysis* 30 (1949): 251–59.

Barshay, Robert Howard. *Philip Wylie: The Man and His Work*. New York: University Press of America, 1979.

Bateson, Gregory. "Metalogue: What Is an Instinct?" In *Approaches to Animal Communication*, edited by Thomas A. Sebeok and Alexandra Ramsay, 11–30. The Hague: Mouton, 1969.

Bateson, P. P. G. "The Characteristics and Context of Imprinting." *Biological Reviews* 41 (1966): 177–217.

Bateson, P. P. G., and Peter H. Klopfer. "Preface." In *Whither Ethology?*, edited by P. P. G. and Peter H. Klopfer, v–viii. Vol. 8 of *Perspectives in Ethology* (1989).

Beach, Frank A. "The Descent of Instinct." *Psychological Review* 62 (1955): 401–10.

———. "Retrospect and Prospect." In *Sex and Behavior*, edited by Frank A. Beach, 535–69. New York: John Wiley, 1965.

Beauvoir, Simone de. *The Second Sex*. New York: Vintage Books, 1953.

Beer, Colin G. "Was Professor Lehrman an Ethologist?" *Animal Behaviour* 23 (1975): 957–64.

———. "Darwin, Instinct, and Ethology." *Journal of the History of the Behavioral Sciences* 19 (1983): 68–80.

————. "Homology, Analogy, and Ethology." *Human Development* 27 (1984): 297–308.

Belsky, Jay, and Laurence D. Steinberg. "The Effects of Day Care: A Critical Review." *Child Development* 49 (1978): 929–49.

Bendau, Clifford P. *Still Worlds Collide: Philip Wylie and the End of the American Dream.* San Bernardino, CA: Borgo Press, 1980.

Benedek, Therese. *Insight and Personality Adjustment: A Study of the Psychological Effects of War.* New York: Ronald Press, 1946.

————. "The Psychosomatic Implications of the Primary Unit: Mother-Child." *American Journal of Orthopsychiatry* 19 (1949): 642–54.

————. "On the Organization of Psychic Energy: Instincts, Drives and Affects." In *Mid-century Psychiatry: An Overview*, edited by Roy R. Grinker, 60–75. Springfield, IL: Charles C. Thomas, 1953.

————. Review of *The Second Sex*, by Simone de Beauvoir. *Psychoanalytic Quarterly* 22 (1953): 264–67.

Benedek, Thomas G. "A Psychoanalytic Career Begins: Therese F. Benedek, M.D.—A Documentary Biography." *Annual of Psychoanalysis* 7 (1979): 3–15.

Berry, Mary Frances. *The Politics of Parenthood: Child Care, Women's Rights, and the Myth of the Good Mother.* Harmondsworth, UK: Penguin, 1994.

Bettelheim, Bruno. "Schizophrenia as a Reaction to Extreme Situations." *American Journal of Orthopsychiatry* 26 (1956): 507–18.

————. *The Empty Fortress: Infantile Autism and the Birth of the Self.* New York: Free Press, 1967.

Biagioli, Mario and Peter Galison, eds. *Scientific Authorship: Credit and Intellectual Property in Science.* New York: Routledge, 2012.

Birns, Beverly, and Niza Ben-Ner. "Psychoanalysis Constructs Motherhood." In *Different Faces of Motherhood*, edited by Beverley Birns and Dale F. Hay, 47–72. New York: Plenum Press, 1988.

Bloch, Ruth. "American Feminine Ideals in Transition: The Rise of the Moral Mother, 1785–1815." *Feminist Studies* 4 (1978): 100–126.

Blomquist, A. J., and Harry F. Harlow. "The Infant Rhesus Monkey Program at the University of Wisconsin Primate Laboratory." *Proceedings of the Animal Care Panel* 11 (1961): 57–64.

Bloom, Lynn Z. *Doctor Spock: Biography of a Conservative Radical.* Indianapolis, IN: Bobbs-Merrill, 1972.

Blum, Deborah. *Love at Goon Park: Harry Harlow and the Science of Affection.* Cambridge, MA: Perseus, 2002.

Bowlby, John. "The Influence of Early Environment in the Development of Neurosis and Neurotic-Character." *International Journal of Psychoanalysis* 21 (1940): 154–78.

————. "Forty-Four Juvenile Thieves: Their Characters and Home-Life." *International Journal of Psychoanalysis* 25 (1944): 19–53, 107–28.

————. "Psychology and Democracy." *Political Quarterly* 17 (1946): 61–76.

————. "Maternal Care and Mental Health," *Bulletin of the World Health Organization* 3 (1951): 355–534.

————. *Child Care and the Growth of Love*, edited by Margery Fry. Harmondsworth, UK: Pelican, 1953.

————. "The Rediscovery of the Family." 1954. In *Rediscovery of the Family and Other Lectures: Sister Marie Hilda Memorial Lectures 1954–1973*, edited by John Bowlby et al., 1–7. Aberdeen, UK: University of Aberdeen Press, 1981.

————. "An Ethological Approach to Research in Child Development." *British Journal of Medical Psychology* 30 (1957): 230–40.

————. "The Nature of the Child's Tie to His Mother." *International Journal of Psychoanalysis* 39 (1958): 350–73.

————. "A Note on Mother-Child Separation as a Mental Health Hazard." *British Journal of Medical Psychology* 31 (1958): 247–48.

————. "Grief and Mourning in Infancy and Early Childhood." *Psychoanalytic Study of the Child* 15 (1960): 9–52.

————. "Separation Anxiety." *International Journal of Psychoanalysis* 41 (1960): 89–113.

————. "Note on Dr. Max Schur's Comments on Grief and Mourning in Infancy and Early Childhood." *Psychoanalytic Study of the Child* 16 (1961): 206–8.

————. *Attachment*. Vol. 1 of *Attachment and Loss*. New York: Basic Books, 1969.

————. *Separation*. Vol. 2 of *Attachment and Loss*. New York: Basic Books, 1973.

————. "Psychoanalysis as Art and Science." *International Review of Psycho-Analysis* 6 (1979): 3–14.

————. "By Ethology Out of Psycho-analysis: An Experiment in Interbreeding." *Animal Behaviour* 28 (1980): 649–56.

————. "Caring for the Young: Influences on Development." In *Parenthood: A Psychodynamic Perspective*, edited by R. S. Cohen, B. J. Cohler, and S. H. Weissman, 269–84. New York: Guilford Press, 1984.

————. *A Secure Base: Parent-Child Attachment and Healthy Human Development*. New York: Basic Books, 1988.

Bowlby, John, Mary Ainsworth, Mary Boston, and Dina Rosenbluth. "The Effects of Mother-Child Separation: A Follow-up Study." *British Journal of Medical Psychology* 29 (1956): 211–47.

Bowlby, John, Karl Figlio, and Robert M. Young. "An Interview with John Bowlby on the Origins and Reception of His Work." *Free Associations* 6 (1986): 36–64.

Bowlby, John, and James Robertson. "A Two-Year-Old Goes to Hospital." *Proceedings of the Royal Society of Medicine* 46, no. 6 (1953): 425–27.

Bowlby, John, James Robertson, and D. Rosenbluth. "A Two-Year-Old Goes to Hospital." *Psychoanalytic Study of the Child* 7 (1952): 82–94.

Breines, Wini. "Domineering Mothers in the 1950s: Image and Reality." *Women's Studies International Forum* 8 (1985): 601–8.

Bretherton, Inge. "The Origins of Attachment Theory: John Bowlby and Mary Ainsworth." *Developmental Psychology* 28 (1992): 759–75.

Bretherton, Inge, and Mary Main. "Mary Dinsmore Salter Ainsworth (1913–1999)." *American Psychologist* 55 (2000): 1148–49.

Briffault, Robert. *The Mothers: A Study of the Origins of Sentiments and Institutions.* 3 vols. New York: Macmillan, 1927.

Brigandt, Ingo. "The Instinct Concept of the Early Konrad Lorenz." *Journal of the History of Biology* 38 (2005): 571–608.

Bugental, James F. T., ed. "Symposium on Karl Bühler's Contributions to Psychology." *Journal of General Psychology*, 75, no. 2 (1966): 181–219.

Buhle, Mari Jo. *Feminism and Its Discontents: A Century of Struggle with Psychoanalysis.* Cambridge, MA: Harvard University Press, 1998.

Burkhardt, Richard W., Jr. "The Founders of Ethology and the Problem of Human Aggression: A Study in Ethologists' Ecologies." In *The Animal/Human Boundary: Historical Perspectives*, edited by Angela N. H. Creager and William Chester Jordan, 265–304. Rochester, NY: University of Rochester Press, 2002.

———. *Patterns of Behavior: Konrad Lorenz, Niko Tinbergen, and the Founding of Ethology.* Chicago: University of Chicago Press, 2005.

Burnham, Donald L. "Freud and Female Sexuality: A Previously Unpublished Letter." *Psychiatry* 34 (August 1971): 328–29.

Byng-Hall, John. "An Appreciation of John Bowlby: His Significance for Family Therapy." *Journal of Family Therapy* 13 (1991): 5–16.

Capshew, James H. *Psychologists on the March: Science, Practice, and Professional Identity in America, 1929–1969.* Cambridge, MA: Cambridge University Press, 1999.

Cartwright, Lisa. "'Emergencies of Survival': Moral Spectatorship and the 'New Vision of the Child' in Postwar Child Psychoanalysis." *Journal of Visual Culture* 3 (2004): 35–49.

Casler, Lawrence. "Maternal Deprivation: A Critical Review of the Literature." *Society for Research in Child Development* 26 (1961): 1–63.

———. "Perceptual Deprivation in Institutional Settings." In *Early Experience and Behavior*, edited by Grant Newton and Seymour Levine, 573–626. Springfield, IL: Charles C. Thomas, 1968.

Chafe, William H. *The Paradox of Change: American Women in the 20th Century.* New York: Oxford University Press, 1991.

Clarke, Ann M., and A. D. B. Clarke. *Early Experience: Myth and Evidence.* New York: Free Press, 1976.

Colin, Virginia L. *Human Attachment.* Philadelphia: Temple University Press, 1996.

Cott, Nancy F. *The Grounding of Modern Feminism.* New Haven, CT: Yale University Press, 1987.

———. *Public Vows: A History of Marriage and the Nation.* Cambridge, MA: Harvard University Press, 2000.

Craig, Wallace. "Appetites and Aversions as Constituents of Instincts." *Biological Bulletin* 34 (1918): 91–107.

Crandall, Floyd M. "Hospitalism." *Archives of Pediatrics* 14 (1897): 448–54.

Cravens, Hamilton. *The Triumph of Evolution: The Heredity-Environment Controversy, 1900–1941*. Baltimore: Johns Hopkins University Press, 1988.

Daston, Lorraine, and Peter Galison. *Objectivity*. Cambridge, MA: MIT Press, 2007.

Daston, Lorraine, and Elizabeth Lunbeck, eds. *Histories of Scientific Observation*. Chicago: University of Chicago Press, 2011.

Daston, Lorraine, and Gregg Mitman, eds. *Thinking with Animals: New Perspectives on Anthropomorphism*. New York: Columbia University Press, 2005.

Daston, Lorraine, and Fernando Vidal, eds. *The Moral Authority of Nature*. Chicago: University of Chicago Press, 2004.

Degler, Carl N. *At Odds: Women and the Family in America from the Revolution to the Present*. New York: Oxford University Press, 1980.

———. *In Search of Human Nature: The Decline and Revival of Darwinism in American Social Thought*. New York: Oxford University Press, 1991.

Deichmann, Ute. *Biologists under Hitler*. Cambridge, MA: Harvard University Press, 1996.

Deutsch, Helene. *Motherhood*. Vol. 2 of *The Psychology of Women: A Psychoanalytic Interpretation*. New York: Grune and Stratton, 1945.

Dewsbury, Donald A. *Comparative Psychology in the Twentieth Century*. Stroudsburg, PA: Hutchinson Ross, 1984.

———, ed. *Studying Animal Behavior: Autobiographies of the Founders*. 1985. Reprint Chicago: University of Chicago Press, 1989.

Dixon, Thomas. *From Passions to Emotions: The Creation of a Secular Psychological Category*. Cambridge: Cambridge University Press, 2003.

Douglas, Susan J., and Meredith W. Michaels. *The Mommy Myth: The Idealization of Motherhood and How It Has Undermined All Women*. New York: Free Press, 2004.

Durant, John R. "Innate Character in Animals and Man: A Perspective on the Origins of Ethology." In *Biology, Medicine and Society, 1840–1940*, edited by Charles Webster, 157–92. Cambridge: Cambridge University Press, 1981.

Durbin, E. F. M., and John Bowlby. *Personal Aggressiveness and War*. New York: Columbia University Press, 1939.

Edwards, Paul N. *The Closed World: Computers and the Politics of Discourse in Cold War America*. Cambridge, MA: MIT Press, 1996.

Ehrenreich, Barbara, and Deirdre English. *For Her Own Good: 150 Years of the Experts' Advice to Women*. New York: Anchor Books, 1978.

Eibl-Eibesfeldt, Irenäus. *Love and Hate: The Natural History of Behavior Patterns*. New York: Holt, Rinehart and Winston, 1972.

Employed Mothers and Child Care. Bulletin of the Women's Bureau 246. Washington, DC: GPO, 1953.

Erikson, Erik H. *Childhood and Society*. New York: Norton, 1950.

———. "Growth and Crises of the 'Healthy Personality.'" In *Symposium on the Healthy Personality,* edited by Milton J. E. Senn, 91–146. New York: Josiah Macy Jr. Foundation, 1950.

———. "Growth and Crises of the 'Healthy Personality.'" In *Personality in Nature, Society and Culture*, edited by Clyde Kluckhohn and Henry A. Murray, 185–225. New York: Knopf, 1953.

———. *Young Man Luther*. London: Faber and Faber, 1958.

Eyer, Diane E. *Mother-Infant Bonding: A Scientific Fiction*. New Haven, CT: Yale University Press, 1992.

———. *Motherguilt: How Our Culture Blames Mothers for What's Wrong with Society*. New York: Random House, 1996.

Feldstein, Ruth. *Motherhood in Black and White: Race and Sex in American Liberalism, 1930–1965*. Ithaca, NY: Cornell University Press, 2000.

Fletcher, Ronald. *Instinct in Man*. Reprint, New York: Schocken Books, 1966.

Föger, Benedikt, and Klaus Taschwer. *Die andere Seite des Spiegels: Konrad Lorenz und der Nationalsozialismus.* Vienna: Czernin, 2001.

Ford Foundation. *Report of the Study for the Ford Foundation on Policy and Program*. Detroit: Ford Foundation, 1949.

Freud, Anna. *The Ego and the Mechanisms of Defense*. London: Hogarth, 1936.

———. "Discussion of Dr. John Bowlby's Paper." *Psychoanalytic Study of the Child* 15 (1960): 53–62.

———. "The Concept of the Rejecting Mother." 1954. In *The Writings of Anna Freud*, vol. 4, *Indications for Child Analysis and Other Papers, 1945–1956,* 586–602. New York: International Universities Press, 1968.

———. "Discussion of John Bowlby's Work on Separation, Grief, and Mourning." In *The Writings of Anna Freud*, vol. 5. *Research at the Hampstead Child-Therapy Clinic and Other Papers, 1956–1965*, 167–86. New York: International Universities Press, 1969.

Freud, Anna, and Dorothy T. Burlingham. *Infants without Families: Reports on the Hampstead Nurseries*. 1943. Reprint in *The Writings of Anna Freud,* vol. 3. New York: International Universities Press, 1973.

———. *War and Children: A Message to American Parents*. 2nd ed. New York: International University Press, 1944.

Freud, Anna, and Sophie Dann. "An Experiment in Group Upbringing." *Psychoanalytic Study of the Child* 6 (1951): 127–68.

Freud, Sigmund. "On Narcissism: An Introduction." 1914. Reprinted in *Standard Edition*, 14:67–102.

———. "Instincts and Their Vicissitudes." 1915. In *Standard Edition*, 14:117–40.

———. "Some Psychical Consequences of the Anatomical Distinction between the Sexes." 1925. In *Standard Edition*, 19:248–58.

———. "Femininity." 1933. In *Standard Edition*, 22:112–35.

———. *An Outline of Psychoanalysis*. 1938. In *Standard Edition*, 23:144–207.

———. *Die Traumdeutung*. In *Gesammelte Werke,* vols. 2–3. London: Imago, 1948.

———. *The Standard Edition of the Complete Psychological Works of Sigmund Freud*. Translated under the general editorship of James Strachey in collaboration with Anna Freud and assisted by Alix Strachey and Alan Tyson. 24 vols. London: Hogarth, 1953–74.

Friedman, Lawrence. *Identity's Architect: A Biography of Erik H. Erikson*. New York: Scribner, 1999.

Fromm-Reichman, Frieda. "Notes on the Development of Treatment of Schizophrenics by Psychoanalytic Psychotherapy." *Psychiatry* 11 (1948): 263–73.

Galison, Peter. *Image and Logic: A Material Culture of Microphysics*. Chicago: University of Chicago Press, 1997.

———. "Ten Problems in History and Philosophy of Science." *ISIS* 99 (2008): 111–24.

Gaskill, Herbert S. "In Memoriam: René A. Spitz, 1887–1974." *Psychoanalytic Study of the Child* 31 (1976): 1–3.

Gay, Peter. *Freud: A Life for Our Time*. New York: Anchor Books, 1988.

Gerall, A. A. "The Effect of Prenatal and Postnatal Injections of Testosterone on Prepubertal Male Guinea Pig Sexual Behavior." *Journal of Comparative and Physiological Psychology* 68 (1963): 92–95.

Gilbert, James. *A Cycle of Outrage: America's Reaction to the Juvenile Delinquent in the 1950s*. New York: Oxford University Press, 1986.

Glueck, Sheldon, and Eleanor Glueck. *Unraveling Juvenile Delinquency*. New York: Commonwealth Fund, 1950.

Goldfarb, William. "Infant Rearing and Problem Behavior." *American Journal of Orthopsychiatry* 13 (1943): 249–65.

———. "Psychological Privation in Infancy and Subsequent Adjustment." *American Journal of Orthopsychiatry* 15 (1945): 247–55.

Gorer, Geoffrey. *The American People: A Study in National Character*. New York: Norton, 1948.

Grant, Julia. *Raising Baby by the Book: The Education of American Mothers*. New Haven, CT: Yale University Press, 1998.

Griffiths, Paul E. "Instinct in the '50s: The British Reception of Konrad Lorenz's Theory of Instinctive Behavior." *Biology and Philosophy* 19 (2004): 609–31.

Griswold, Robert L. *Fatherhood in America: A History*. New York: Basic Books, 1993.

Grob, Gerald N. *The Mad among Us: A History of the Care of America's Mentally Ill*. Cambridge, MA: Harvard University Press, 1994.

Gross, Daniel M. *The Secret History of Emotion: From Aristotle's Rhetoric to Modern Brain Science*. Chicago: University of Chicago Press, 2006.

Haldane, J. B. S. "The Sources of Some Ethological Notions." *British Journal of Animal Behaviour* 4 (1956): 162–64.

Hale, Nathan G., Jr. *The Rise and Crisis of Psychoanalysis in the United States: Freud and the Americans, 1917–1985.* New York: Oxford University Press, 1995.

Haraway, Donna. *Primate Visions: Gender, Race, and Nature in the World of Modern Science.* New York: Routledge, 1989.

Hardyment, Christina. *Dream Babies: Three Centuries of Good Advice on Child Care.* New York: Harper and Row, 1983.

Harlow, Harry F. "Mice, Monkeys, Men, and Motives." *Psychological Review* 60 (1953): 23–32.

———. "The Nature of Love." *American Psychologist* 13 (1958): 673–85.

———. "Basic Social Capacities of Primates." *Human Biology* 31 (1959): 40–53.

———. "Love in Infant Monkeys." *Scientific American* 200, no. 6 (1959): 68–74.

———. "Affectional Behavior in the Infant Monkey." In *The Central Nervous System and Behavior: Transactions of the Third Conference, February 21–24, 1960,* edited by Mary A. B. Brazier, 307–57. New York: Josiah Macy Jr. Foundation, 1960.

———. "Of Love in Infants: Studies of Monkeys Show How Affection for Mothers Is Formed." *Natural History* 69, no. 5 (1960): 18–23.

———. "The Development of Affectional Patterns in Infant Monkeys." In *Determinants of Infant Behaviour: Proceedings of the Tavistock Seminar on Mother-Infant Interaction,* vol. 1, edited by Brian M. Foss, 75–97. London: Methuen, 1961.

———. "Affectional Systems of Monkeys, Involving Relations between Mothers and Young." In *International Symposium on Comparative Medicine Proceedings,* 6–10. New York: Eaton Laboratories, 1962.

———. "Motivation in Monkeys—and Men." In *Psychology and Life,* 6th ed., edited by Floyd Leon Ruch, 589–94. Chicago: Scott, Foresman, 1963.

———. "A Behavioral Approach to Psychoanalytic Theory." *Science and Psychoanalysis* 7 (1964): 93–113.

———. "Early Social Deprivation and Later Behavior in the Monkey." In *Unfinished Tasks in the Behavioral Sciences,* edited by Arnold Abrams, Harry H. Garner, and James E. P. Toman, 154–73. Baltimore: Williams and Wilkins, 1964.

———. "The Primate Socialization Motives." *Transactions and Studies of the College of Physicians of Philadelphia,* 4th ser., 33 (1966): 224–37.

———. "Monkeys, Men, Mice, Motives, and Sex." In *Psychological Research: The Inside Story,* edited by Michael H. Siegel and H. Philip Zeigler, 3–22. New York: Harper and Row, 1976.

———. "Birth of the Surrogate Mother." In *Discovery Processes in Modern Biology,* edited by W. R. Klemm, 134–50. Huntington, NY: Robert E. Krieger, 1977.

Harlow, Harry F., and Margaret K. Harlow. "A Study of Animal Affection." *Natural History* 70 (1961): 48–55.

————. "The Effect of Rearing Conditions on Behavior." *Bulletin of the Men-ninger Clinic* 26 (1962): 213–24.

————. "The Affectional Systems." In *Behavior of Nonhuman Primates*, vol. 2, edited by H. F. Harlow, A. M. Schrier, and F. Stollnitz, 287–334. New York: Academic Press, 1965.

————. "Effects of Various Mother-Infant Relationships on Rhesus Monkey Behaviors." In *Determinants of Infant Behaviour: Proceedings of the Tavistock Seminar on Mother-Infant Interaction*, vol. 4, edited by Brian M. Foss, 15–36. London: Methuen, 1969.

————. "Developmental Aspects of Emotional Behavior." In *Psychological Correlates of Emotion*, edited by P. Black, 37–58. New York: Academic Press, 1970.

Harlow, Harry F., M. K. Harlow, R. O. Dodsworth, and G. L. Arling. "Maternal Behavior of Rhesus Monkeys Deprived of Mothering and Peer Association in Infancy." *Proceedings of the Philosophical Society* 110 (1966): 58–66.

Harlow, Harry F., and B. Seay. "Affectional Systems in Rhesus Monkeys." *Journal of the Arkansas Medical Society* 61, no. 4 (1964): 107–10.

Harlow Harry F., and R. Stagner. "Psychology of Feelings and Emotions: I. Theory of Feelings." *Psychological Review* 39 (1932): 570–89.

————. "Psychology of Feelings and Emotions: II. Theory of Emotions." *Psychological Review* 40 (1932): 184–95.

Harlow, Harry F., and R. R. Zimmerman. "The Development of Affectional Responses in Infant Monkeys." *Proceedings of the American Philosophical Society* 102, no. 5 (1958): 501–9.

————. "Affectional Responses in the Infant Monkey." *Science* 130 (1959): 421–32.

Harrington, Anne. *Reenchanted Science: Holism in German Culture from Wilhelm II to Hitler.* Princeton, NJ: Princeton University Press, 1996.

Hartmann, Heinz. "Comments on the Psychoanalytic Theory of Instinctual Drives." In *Essays in Ego Psychology*, 69–89. London: Hogarth Press and the Institute of Psychoanalysis, 1964.

Hartmann, Susan M. "Prescriptions for Penelope: Literature on Women's Obligations to Returning World War II Veterans." *Women's Studies* 5 (1978): 223–39.

Hawes, Joseph M. *Children between the Wars: American Childhood, 1920–1940.* New York: Twayne, 1997.

Hebb, Donald O. "Heredity and Environment in Mammalian Behaviour." *British Journal of Animal Behaviour* 1 (1953): 43–47.

Herman, Ellen. *The Romance of American Psychology: Political Culture in the Age of Experts.* Berkeley: University of California Press, 1995.

————. "Families Made by Science: Arnold Gesell and the Technologies of Modern Child Adoption." *ISIS* 92 (2001): 684–715.

————. *Kinship by Design: A History of Adoption in the Modern United States.* Chicago: University of Chicago Press, 2008.

Hess, Eckhard H. "Imprinting." *Science* 130 (1959): 133–41.

———. "On Anthropomorphism." In *Unfinished Tasks in the Behavioral Sciences,* edited by Arnold Abrams, Harry H. Garner, and James E. P. Tolman, 174–78. Baltimore: Williams and Wilkins, 1964.

Hinde, Robert A. "The Modifiability of Instinctive Behaviour." *Advancement of Science* 12 (1955): 19–24.

———. "Ethological Models and the Concept of 'Drive.' " *British Journal for the Philosophy of Science* 6 (1956): 321–31.

———. "Unitary Drives." *Animal Behaviour* 7 (1959): 130–40.

———. "Energy Models of Motivation." 1960. Reprinted in *Foundations of Animal Behavior: Classic Papers with Commentaries,* edited by Lynne D. Houck and Lee C. Drickamer, 513–27. Chicago: University of Chicago Press, 1996.

———. "Some Aspects of the Imprinting Problem." *Symposia of the Zoological Society of London* 8 (1962): 129–38.

———. "The Nature of Imprinting." In *Determinants of Infant Behaviour: Proceedings of the Tavistock Seminar on Mother-Infant Interaction,* vol. 2, edited by B. M. Foss, 227–33. London: Methuen, 1963.

———. *Biological Bases of Human Social Behavior.* New York: McGraw-Hill, 1974.

———. "Mother-Infant Separation and the Nature of Inter-individual Relationships: Experiments with Rhesus Monkeys." *Proceedings of the Royal Society,* ser. B, 196 (1977): 29–50.

———. "Attachment: Some Conceptual and Biological Issues." In *The Place of Attachment in Human Behavior,* edited by Colin Murray Parkes and Joan Stevenson-Hinde, 60–76. New York: Basic Books, 1982.

———. "Relationships, Attachment, and Culture: A Tribute to John Bowlby." *Infant Mental Health Journal* 12 (1991): 154–63.

———. "Konrad Lorenz and Nikolaas Tinbergen." In *Seven Pioneers of Psychology,* edited by R. Fuller, 74–105. London: Routledge, 1995.

———. "Ethology and Attachment Theory." In *Attachment from Infancy to Adulthood: The Major Longitudinal Studies,* edited by Klaus Grossmann, Karin Grossmann, and Everett Waters, 1–12. New York: Guilford, 2005.

Hinde, Robert A., and Lynda Davies. "Removing Infant Rhesus from Mother for 13 Days Compared with Removing Mother from Infant." *Journal of Child Psychology and Psychiatry* 13 (1972): 227–37.

Hinde, Robert A., and L. McGinnis. "Some Factors Influencing the Effects of Temporary Mother-Infant Separation: Some Experiments with Rhesus Monkeys." *Psychological Medicine* 7 (1977): 197–222.

Hinde, Robert A., and Joan Stevenson-Hinde. "Attachment: Biological, Cultural and Individual Desiderata." *Human Development* 33 (1990): 62–72.

Hochschild, Arlie Russell. *The Managed Heart.* Berkeley: University of California Press, 2003.

Hofer, Veronika. "Konrad Lorenz als Schüler von Karl Bühler: Diskussion der neu entdeckten Quellen zu den persönlichen und inhaltlichen Positionen zwischen Karl Bühler, Konrad Lorenz und Egon Brunswik." *Die Zeitgeschichte* 28 (2001): 135–59.

Hoffman, Howard S., and Alan M. Ratner. "A Reinforcement Model of Imprinting: Implications for Socialization in Monkeys and Men." *Psychological Review* 80 (1973): 527–44.

Hollingsworth, Leta S. "Social Devices for Impelling Women to Bear and Rear Children." *American Journal of Sociology* 22 (1916): 19–29.

Holmes, Jeremy. *John Bowlby and Attachment Theory.* London: Routledge, 1993.

Horney, Karen. "Premenstrual Tension." In *Feminine Psychology*, 99–106. 1931. Reprint New York: Norton, 1967.

Howells, J. G. "Fallacies in Child Care. 1. That 'Separation Is Synonymous with Deprivation.'" *Acta Paedopsychiatrica* 37 (1970): 3–14.

Hrdy, Sarah Blaffer. *Mother Nature: Maternal Instincts and the Shaping of the Species.* London: Vintage Books, 2000.

Hulbert, Ann. *Raising America: Experts, Parents, and a Century of Advice about Children.* New York: Vintage Books, 2003.

Hunter, Virginia. "John Bowlby: An Interview." *Psychoanalytic Review* 78 (1991): 159–75.

James, William. *Principles of Psychology.* 1890. Reprint New York: Holt, 1900.

Jaynes, Julian. "Imprinting: The Interaction of Learned and Innate Behavior: I. Development and Generalization." *Journal of Comparative and Physiological Psychology* 49 (1956): 201–6.

———. "Imprinting: The Interaction of Learned and Innate Behavior: II. The Critical Period." *Journal of Comparative and Physiological Psychology* 50 (1957): 6–10.

———. "Imprinting: The Interaction of Learned and Innate Behavior: III. Practice Effects on Performance, Retention, and Fear." *Journal of Comparative and Physiological Psychology* 51 (1958): 234–37.

———. "Imprinting: The Interaction of Learned and Innate Behavior: IV. Generalization and Emergent Discrimination." *Journal of Comparative and Physiological Psychology* 51 (1958): 238–42.

Jones, Kathleen W. *Taming the Troublesome Child: American Families, Child Guidance, and the Limits of Psychiatric Authority.* Cambridge, MA: Harvard University Press, 1999.

Jordan, John M. *Machine-Age Ideology: Social Engineering and American Liberalism, 1911–1939.* Chapel Hill: University of North Carolina Press, 1994.

Kagan, Jerome. *Unstable Ideas: Temperament, Cognition, and Self.* Cambridge, MA: Harvard University Press, 1989.

Kalikow, Theodora J. "History of Konrad Lorenz's Ethological Theory, 1927–1939: The Role of Meta-theory, Theory, Anomaly and New Discoveries in a Scientific 'Evolution.'" *Studies in the History and Philosophy of Science* 6 (1975): 331–41.

———. "Konrad Lorenz's Ethological Theory, 1939–1943: 'Explanations' of

Human Thinking, Feeling, and Behavior." *Philosophy of the Social Sciences* 6 (1976): 15–34.

Kanner, Leo. "Autistic Disturbances of Affective Contact." *Nervous Child* 2 (1943): 217–50.

Karen, Robert. *Becoming Attached: First Relationships and How They Shape Our Capacity to Love.* New York: Oxford University Press, 1998.

Kaufman, I. Charles. "Biologic Considerations of Parenthood." In *Parenthood: Its Psychology and Psychopathology*, edited by E. James Anthony and Therese Benedek, 3–55. 1970. Reprint Boston: Little, Brown, 1996.

———. "Mother/Infant Relations in Monkeys and Humans: A Reply to Professor Hinde." In *Ethology and Psychiatry*, edited by Norman F. White, 47–68. Toronto: University of Toronto Press, 1974.

Keefer, Truman Frederick. *Philip Wylie.* Boston: Twayne, 1977.

Keller, Evelyn Fox. "Physics and the Emergence of Molecular Biology: A History of Cognitive and Political Synergy." *Journal of the History of Biology* 23 (1990): 389–409.

———. *Refiguring Life: Metaphors of Twentieth-Century Biology.* New York: Columbia University Press, 1995.

———. *The Mirage of a Space between Nature and Nurture.* Durham, NC: Duke University Press, 2010.

Kennedy, J. S. "Is Modern Ethology Objective?" *British Journal of Animal Behaviour* 2 (1954): 12–19.

Kinsey, Alfred C., Wardell B. Pomeroy, Clyde E. Martin, and Paul H. Gebhard. *Sexual Behavior in the Human Female.* Philadelphia: W. B. Saunders, 1953.

Klopfer, Peter H. *Politics and People in Ethology: Personal Reflections on the Study of Animal Behavior.* Lewisburg, PA: Bucknell University Press, 1999.

Kohler, Robert. *Lords of the Fly:* Drosophila *Genetics and the Experimental Life.* Chicago: University of Chicago Press, 1994.

Komarovsky, Mirra. "Cultural Contradictions and Sex Roles." *American Journal of Sociology* 52 (1946): 184–89.

———. "Functional Analysis of Sex Roles." *American Sociological Review* 15 (1950): 508–16.

———. *Women in the Modern World: Their Education and Their Dilemmas.* Boston: Little, Brown, 1953.

Kruuk, Hans. *Niko's Nature: A Life of Niko Tinbergen and His Science of Animal Behavior.* Oxford: Oxford University Press, 2003.

Ladd-Taylor, Molly, and Lauri Umansky, eds. *"Bad" Mothers: The Politics of Blame in Twentieth-Century America.* New York: New York University Press, 1998.

Lamb, Michael E., Ross A. Thompson, William P. Gardner, Eric L. Charnov, and David Estes. "Security of Infantile Attachment as Assessed in the 'Strange Situation': Its Study and Biological Interpretation." *Behavioral and Brain Sciences* 7 (1984): 127–47.

Lebovici, Serge. "The Concept of Maternal Deprivation: A Review of Research."

In World Health Organization, *Deprivation of Maternal Care: A Reassessment of Its Effects,* 75–96. Geneva: World Health Organization, 1962.

Lehrman, Daniel S. "Comparative Behavior Studies." *Bird-Banding* 12 (1941): 86–87.

———. "A Critique of Konrad Lorenz's Theory of Instinctive Behavior." *Quarterly Review of Biology* 28 (1953): 337–63.

———. "The Physiological Basis of Parental Feeding Behavior in the Ring Dove (*Streptopelia risoria*)." *Behaviour* 7 (1955): 241–86.

———. "On the Organization of Maternal Behavior and the Problem of Instinct." In *L'instinct dans le comportement des animaux et de l'homme,* edited by Mario Autuori et al., 475–520. Paris: Masson, 1956.

———. "On the Origin of the Reproductive Cycle in Doves." *Transactions of the New York Academy of Sciences* 21 (1959): 682–88.

———. "Hormonal Regulation of Parental Behavior in Birds and Infrahuman Mammals." In *Sex and Internal Secretions,* edited by W. C. Young, 1268–82. Baltimore: Williams and Wilkins, 1961.

———. "Ethology and Psychology," *Recent Advances in Biological Psychiatry* 4 (1963): 86–94.

———. "Interaction between Internal and External Environments in the Regulation of the Reproductive Cycle of the Ring Dove." In *Sex and Behavior,* edited by Frank A. Beach, 355–80. New York: John Wiley, 1965.

———. "Semantic and Conceptual Issues in the Nature-Nurture Problem." In *Development and Evolution of Behavior: Essays in Memory of T. C. Schneirla,* edited by L. R. Aronson, E. Tobach, J. S. Rosenblatt, and D. S. Lehrman, 17–52. San Francisco: Freeman, 1970.

———. "Can Psychiatrists Use Ethology?" In *Ethology and Psychiatry,* edited by Norman F. White, 187–96. Toronto: University of Toronto Press, 1974.

Lehrman, Daniel S., and Rochelle P. Wortis. "Previous Breeding Experience and Hormone-Induced Incubation Behavior in the Ring Dove." *Science* 132 (1960): 1667–68.

LeRoy, H. A. "Harry Harlow: From the Other Side of the Desk." *Integrative Psychological and Behavioral Science* 42, no. 4 (2008): 348–53.

LeRoy, H. A., and G. A. Kimble. "Harry Frederick Harlow: And One Thing Led to Another. . . ." In *Portraits of Pioneers in Psychology,* vol. 5, edited by Gregory A. Kimble and Michael Wertheimer, 279–97. Washington, DC: American Psychological Association, 2003.

Levy, David M. "Primary Affect Hunger." *American Journal of Psychiatry* 94 (1937): 643–52.

———. "Psychopathic Personality and Crime." *Journal of Educational Sociology* 16 (1942): 99–115.

———. "Psychosomatic Studies of Some Aspects of Maternal Behavior." 1942. Reprint in *Personality in Nature, Society, and Culture,* edited by Clyde Kluckhohn and Henry A. Murray, 104–10. New York: Knopf, 1953.

———. *Maternal Overprotection.* 1943. Reprint New York: Norton, 1966.

Lewis, Jan. "Mother's Love: The Construction of an Emotion in Nineteenth-Century America." In *Mothers and Motherhood: Readings in American History,* edited by Rima D. Apple and Janet Golden, 52–71. Columbus: Ohio State University Press, 1997.

Lewis, Michael. *Altering Fate: Why the Past Does Not Predict the Future.* New York: Guilford Press, 1997.

Lewis, Thomas, Fari Amini, and Richard Lannon. *A General Theory of Love.* New York: Vintage, 2000.

Leys, Ruth. "The Turn to Affect: A Critique." *Critical Inquiry* 37 (2011): 434–72.

Lorenz, Konrad. "Beobachtungen an Dohlen." *Journal für Ornithologie* 75 (1927): 511–19.

———. "Beiträge zur Ethologie sozialer Corviden." *Journal für Ornithologie* 79 (1931): 67–127.

———. "A Consideration of Methods of Identification of Species-Specific Instinctive Behaviour Patterns in Birds." In Konrad Lorenz, *Studies in Animal and Human Behavior,* 1:57–100. Cambridge, MA: Harvard University Press, 1970–71. Originally published as "Betrachtungen über das Erkennen der arteigenen Triebhandlungen der Vögel." *Journal für Ornithologie* 80 (1932): 50–98.

———. "Beobachtungen an freifliegenden zahmgehaltenen Nachtreihern." *Journal für Ornithologie* 82 (1934): 160–61.

———. "Companions as Factors in the Bird's Environment: The Conspecific as the Eliciting Factor for Social Behaviour Patterns." In Konrad Lorenz, *Studies in Animal and Human Behavior,* 1:101–258. Cambridge, MA: Harvard University Press, 1970–71. Originally published as "Der Kumpan in der Umwelt des Vogels, der Artgenosse als auslösendes Moment sozialer Verhaltensweisen." *Journal für Ornithologie* 83 (1935): 137–215, 289–413.

———. "The Establishment of the Instinct Concept." In *Studies in Animal and Human Behavior,* 1:259–315. Cambridge, MA: Harvard University Press, 1970–71. Originally published as "Über die Bildung des Instinktbegriffes." *Die Naturwissenschaften* 25 (1937): 289–300, 307–18, 324–31.

———. "The Nature of Instinct." 1937. In *Instinctive Behavior: The Development of a Modern Concept,* edited by Claire H. Schiller, 129–75. New York: International Universities Press, 1957.

———. "Über Ausfallserscheinungen im Instinktverhalten von Haustieren und ihre sozialpsychologische Bedeutung." In *Charakter und Erziehung: Bericht über den 16. Kongress der deutschen Gesellschaft für Psychologie in Bayreuth,* edited by Otto Klemm, 139–47. Leipzig: Teubner, 1939.

———. "Durch Domestikation verursachte Störungen arteigenen Verhaltens." *Zeitschrift für angewandte Psychologie und Charakterkunde* 59, nos. 1–2 (1940): 2–81.

———. "Die angeborenen Formen möglicher Erfahrung." *Zeitschrift für Tierpsychologie* 5 (1943): 235–409.

————. *King Solomon's Ring: New Light on Animal Ways.* New York: Crowell, 1952. Originally published as *Er redete mit dem Vieh, den Vögeln und den Fischen.* Vienna: Borotha-Schoeler, 1949.

————. "The Comparative Method in Studying Innate Behaviour Patterns." *Symposia of the Society for Experimental Biology* 4 (1950): 221–68.

————. "The Past Twelve Years in the Comparative Study of Behavior." 1952. Reprint in *Instinctive Behavior: The Development of a Modern Concept,* edited by Claire H. Schiller, 288–310. New York: International Universities Press, 1957.

————. "Über angeborene Instinktformeln beim Menschen." *Deutsche medizinische Wochenschrift* 78 (1953): 1566–69, 1600–1604.

————. "Psychology and Phylogeny." 1954. In *Studies in Animal and Human Behavior,* 2:196–245. Cambridge, MA: Harvard University Press, 1970–71.

————. "The Objectivistic Theory of Instinct." In *L'instinct dans le comportement des animaux et de l'homme,* edited by Mario Autuori et al., 51–76. Paris: Masson, 1956.

————. "Companionship in Bird Life: Fellow Members of the Species as Releasers of Social Behavior." In *Instinctive Behavior: The Development of a Modern Concept,* edited by Claire H. Schiller, 83–128. New York: International Universities Press, 1957.

————. *Evolution and Modification of Behavior.* Chicago: University of Chicago Press, 1965.

————. *On Aggression.* New York: Harcourt Brace and World, 1966.

————. *Studies in Animal and Human Behaviour.* 2 vols. Cambridge, MA: Harvard University Press, 1970–71.

————. *Civilized Man's Eight Deadly Sins.* New York: Harcourt, 1973.

————. "The Fashionable Fallacy of Dispensing with Description." *Die Naturwissenschaften* 60 (1973): 1–9.

————. "Analogy as a Source of Knowledge." *Science* 185 (1974): 229–34.

————. *The Waning of Humaneness.* Boston: Little, Brown, 1987.

————. *Here Am I — Where Are You? The Behavior of the Greylag Goose.* New York: Harcourt Brace Jovanovich, 1991.

————. *The Natural Science of the Human Species: An Introduction to Comparative Behavioral Research. The "Russian Manuscript" (1944–1948).* Cambridge, MA: MIT Press, 1996.

Lorenz, Konrad, and Niko Tinbergen, "Taxis and Instinctive Behaviour Pattern in Egg-Rolling by the Greylag Goose." In Lorenz, *Studies in Animal and Human Behaviour,* 1:316–50. Originally published as "Taxis und Instinkthandlung in der Eirollbewegung der Graugans." *Zeitschrift für Tierpsychologie* 2 (1938): 1–29.

Lunbeck, Elizabeth. *The Psychiatric Persuasion: Knowledge, Gender, and Power in Modern America.* Princeton, NJ: Princeton University Press, 1994.

Lundberg, Ferdinand, and Marynia F. Farnham, *Modern Woman: The Lost Sex.* New York: Harper, 1947.

Maier, Thomas. *Dr. Spock: An American Life.* New York: Harcourt Brace, 1998.

Margolis, Maxine L. *Mothers and Such: Views of American Women and Why They Changed*. Berkeley: University of California Press, 1984.

May, Elaine Tyler. *Homeward Bound: American Families in the Cold War Era*. New York: Basic Books, 1988.

Mayr, Ernst. *Populations, Species, and Evolution*. Cambridge, MA: Harvard University Press, 1963.

McDougall, William. *An Introduction to Social* Psychology. 1908. Reprint Boston: Luce, 1916.

McGregor, Gaile. "Domestic Blitz: A Revisionist Theory of the Fifties." *American Studies* 34 (1993): 5–33.

McKinney, William T., Jr., Stephen J. Suomi, and Harry F. Harlow. "Studies in Depression." *Psychology Today* 4, no. 12 (1971): 61–63.

Mead, Margaret. *And Keep Your Powder Dry: An Anthropologist Looks at America*. New York: William Morrow, 1943.

———. "A Cultural Anthropologist's Approach to Maternal Deprivation." In World Health Organization, *Deprivation of Maternal Care: A Reassessment of Its Effects*, 45–62. Geneva: World Health Organization, 1962.

Melosh, Barbara. *Strangers and Kin: The American Way of Adoption*. Cambridge, MA: Harvard University Press, 2002.

Menninger, Karl A., with the collaboration of Jeanetta Lyle Menninger, 1942. *Love against Hate*. New York: Harcourt, Brace, 1942.

Menninger, William C. *Psychiatry in a Troubled World: Yesterday's War and Today's Challenge*. New York: Macmillan, 1948.

Meyerowitz, Joanne. "Beyond the Feminine Mystique: A Reassessment of Postwar Mass Culture, 1946–1958." *Journal of American History* 79 (1993): 1455–82.

———, ed. *Not June Cleaver: Women and Gender in Postwar America, 1945–1960*. Philadelphia: Temple University Press, 1994.

Michel, Sonya. *Children's Interests/Mothers' Rights: The Shaping of America's Child Care Policy*. New Haven, CT: Yale University Press, 1999.

Midgley, Nick. "Anna Freud: The Hampstead War Nurseries and the Role of the Direct Observations of Children in Psychoanalysis." *International Journal of Psychoanalysis* 88 (2007): 939–59.

Mintz, Steven, and Susan Kellogg. *Domestic Revolutions: A Social History of American Family Life*. New York: Free Press, 1988.

Mitchell, Robert W., Nicholas S. Thompson, and H. Lyn Miles, eds. *Anthropomorphism, Anecdotes, and Animals*. Albany: State University of New York Press, 1997.

Mitman, Gregg. *Reel Nature: America's Romance with Wildlife on Film*. Cambridge, MA: Harvard University Press, 1999.

Mitman, Gregg, and Richard W. Burkhardt. "Struggling for Identity: The Study of Animal Behavior in America, 1930–1945." In *The Expansion of American Biology*, edited by Keith R. Benson, Jane Maienschein, and Ronald Rainger, 164–94. New Brunswick, NJ: Rutgers University Press, 1991.

Moltz, Howard. "Imprinting: Empirical Basis and Theoretical Significance." *Psychological Bulletin* 57 (1960): 291–314.

Montagu, Ashley, ed. *Man and Aggression*. New York: Oxford University Press, 1968.

———. *The Natural Superiority of Women*. New York: Macmillan, 1953.

———. "The Origins and Meaning of Love." In *The Meaning of Love*, edited by Ashley Montagu. New York: Julian Press, 1953.

———. "The Triumph and Tragedy of the American Woman." *Saturday Review* 41 (September 27, 1958): 13–15, 34–35.

Morgan, Elaine. *The Descent of the Child: Human Evolution from a New Perspective*. London: Penguin, 1996.

Morss, John R. *The Biologising of Childhood: Developmental Psychology and the Darwinian Myth*. Hove, UK: Lawrence Erlbaum, 1990.

Munz, Tania. "Die Ethologie des wissenschaftlichen Cineasten: Karl von Frisch, Konrad Lorenz und das Verhalten der Tiere im Film." *Montage AV* 14, no. 4 (2005): 52–68.

———. "'My Goose Child Martina': The Multiple Uses of Geese in Konrad Lorenz's Writings on Animals, 1935–1988." *Historical Studies in the Natural Sciences* 41 (2011): 405–56.

Myerson, Abraham. "Let's Quit Blaming Mom." *Science Digest* 29 (1951): 10–15.

Myrdal, Alva, and Viola Klein. *Women's Two Roles: Home and Work*. London: Routledge and Kegan Paul, 1956.

Nisbett, Alec. *Konrad Lorenz*. New York: Harcourt Brace Jovanovich, 1976.

Nussbaum, Martha C. *Upheavals of Thought: The Intelligence of Emotions*. Cambridge: Cambridge University Press, 2001.

O'Connor, Neil. "The Evidence for the Permanently Disturbing Effects of Mother-Child Separation." *Acta Psychologica* 12 (1956): 174–91.

Orlansky, Harold. "Infant Care and Personality." *Psychological Bulletin* 46 (1949): 1–48.

Ostow, Mortimer. "Psychoanalysis and Ethology." *Journal of the American Psychoanalytic Association* 8 (1960): 526–34.

Paul, Diane. *Controlling Human Heredity: 1865 To The Present*. Atlantic Highlands, N.J: Humanities Press, 1995.

Pells, Richard. *The Liberal Mind in a Conservative Age: American Intellectuals in the 1940s and 1950s*. Middletown, CT: Wesleyan University Press, 1985.

Piel, Gerard. "The Comparative Psychology of T. C. Schneirla." In *Development and Evolution of Behavior*, edited by Lester R. Aronson, Ethel Tobach, Daniel S. Lehrman, and Jay S. Rosenblatt, 1–13. San Francisco: Freeman, 1970.

Pinneau, Samuel R. "A Critique of the Articles by Margaret Ribble." *Child Development* 21 (1950): 203–28.

———. "The Infantile Disorders of Hospitalism and Anaclitic Depression." *Psychological Bulletin* 52 (1955): 429–52.

Plant, Rebecca Jo. "The Veteran, His Wife and Their Mothers: Prescriptions for Psychological Rehabilitation after World War II." In *Tales of the Great American Victory: World War II in Politics and Poetics*, edited by Diederik Oost- dijk and Markha G. Valenta, 95–100. Amsterdam: Vrije University Press, 2006.

———. *Mom: The Transformation of Motherhood in Modern America.* Chicago: University of Chicago Press, 2010.

Pratt, George K. *Soldier to Civilian: Problems of Readjustment.* New York: McGraw-Hill, 1944.

Prugh, Dane G., and Robert G. Harlow. "'Masked Deprivation' in Infants and Young Children." In World Health Organization, *Deprivation of Maternal Care: A Reassessment of Its Effects, 9–30.* Geneva: World Health Organization, 1962.

Radick, Gregory. *The Simian Tongue: The Long Debate about Animal Language.* Chicago: University of Chicago Press, 2007.

Rajecki, D. W., Michael E. Lamb, and Pauline Obmascher. "Toward a General Theory of Infantile Attachment: A Comparative Review of Aspects of the Social Bond." *Behavioral and Brain Sciences* 3 (1978): 417–64.

Reddy, William M. "Historical Research on the Self and Emotions." *Emotion Review* 1 (2009): 302–15.

Reed, Ruth. "Changing Conceptions of the Maternal Instinct." *Journal of Abnormal Psychology and Social Psychology* 18 (1923): 78–87.

Ribble, Margaret [Margarethe]. "Disorganizing Factors of Infant Personality." *American Journal of Psychiatry* 98 (1941): 459–63.

———. *The Rights of Infants: Early Psychological Needs and Their Satisfaction.* New York: Columbia University Press, 1943.

———. "The Rights of Infants." *Ladies' Home Journal* 61 (August 1944): 20–21, 135–40.

Richards, Robert. "The Innate and the Learned: the Evolution of Konrad Lorenz's Theory of Instinct." *Philosophy of the Social Sciences* 4 (1974): 111–33.

Ridley, Matt. *Genome: The Autobiography of a Species in 23 Chapters.* New York: HarperCollins, 1999.

Ritvo, Lucille B. *Darwin's Influence on Freud: A Tale of Two Sciences.* New Haven, CT: Yale University Press, 1972.

Robertson, James. *A Two-Year-Old Goes to Hospital.* London: Tavistock Child Development Research Unit, 1952. Filmstrip.

Robertson, James, and John Bowlby. "Responses of Young Children to Separation from Their Mothers: II. Observations of the Sequences of Response of Children Aged 18 to 24 Months during the Course of Separation." *Courrier du Centre International de l'Enfance* 3 (1952): 131–42.

Robertson, James, and Joyce Robertson. "Young Children in Brief Separation: A Fresh Look." *Psychoanalytic Study of the Child* 26 (1971): 264–315.

Rodman, F. Robert. *The Spontaneous Gesture: Selected Letters of D. W. Winnicott.* Cambridge, MA: Harvard University Press, 1987.

Röell, D. R. *The World of Instinct: Niko Tinbergen and the Rise of Ethology in the Netherlands (1920–1950)*. Assen, Netherlands: Van Gorcum, 2000.

Rosenberg, Charles E. "History and Experience." In *The Family in History*, edited by Charles E. Rosenberg, 1–11. Philadelphia: University of Pennsylvania Press, 1975.

Rosenberg, Charles E., and Carroll Smith-Rosenberg. "The Female Animal: Medical and Biological Views of Women." In *No Other Gods: On Science and American Social Thought*, edited by Charles Rosenberg, 54–70. Baltimore: Johns Hopkins University Press, 1976.

Rosenblatt, Jay S. "Daniel Sanford Lehrman, June 1, 1919–August 27, 1972." *Biographical Memoirs of the National Academy of Sciences* 66 (1995): 227–45.

Rosenwein, Barbara H. "Worrying about Emotions in History: Review Essay." *American Historical Review* 107 (2002): 821–45.

Russett, Cynthia Eagle. *Sexual Science: The Victorian Construction of Womanhood*. Cambridge, MA: Harvard University Press, 1989.

Rustin, Michael. "Attachment in Context." In *The Politics of Attachment: Towards a Secure Society*, edited by Sebastian Kraemer and Jane Roberts, 212–28. London: Free Association Books, 1996.

Salter, Mary D. *An Evaluation of Adjustment Based upon the Concept of Security*. Toronto: University of Toronto Press, 1940.

Sayers, Janet. *Mothers of Psychoanalysis: Helene Deutsch, Karen Horney, Anna Freud, and Melanie Klein*. New York: Norton, 1991.

Schaffner, Bertram, ed. *Group Processes: Transactions of the First Conference, September 26–30, 1954, Ithaca, New York*. New York: Josiah Macy Jr. Foundation, 1955.

Schiller, Claire H., ed. *Instinctive Behavior: The Development of a Modern Concept*. New York: International Universities Press, 1957.

Schneirla, T. C. "The Relationship between Observation and Experimentation in the Field Study of Behavior." *Annals of the New York Academy of Sciences* 51 (1950): 1022–44.

———. "A Consideration of Some Conceptual Trends in Comparative Psychology." *Psychological Bulletin* 49 (1952): 559–97.

———. "Behavioral Development and Comparative Psychology." *Quarterly Review of Biology* 41 (1966): 283–302.

———. *Army Ants: A Study in Social Organization*. Edited by Howard R. Topoff. San Francisco, CA: W. H. Freeman, 1971.

Schneirla, T. C., and L. M. Chace. "Carpenter Ants." *Natural History* 60 (May 1951): 227–33.

Schneirla, T. C., and G. Piel. "The Army Ant." *Scientific American* 178 (June 1948): 16–23.

Schneirla, T. C., and Jay S. Rosenblatt. "Behavioral Organization and Genesis of the Social Bond in Insects and Mammals." *American Journal of Orthopsychiatry* 31 (1961): 223–91.

———. "Critical Periods in the Development of Behavior." *Science* 139 (1963): 1110–15.

Schur, Max. "Discussion of Dr. John Bowlby's Paper." *Psychoanalytic Study of the Child* 15 (1960): 63–66.

———. "Discussion: A Psychoanalyst's Comments." *American Journal of Orthopsychiatry* 31 (1961): 276–91.

Schwartz, Joseph. *Cassandra's Daughter: A History of Psychoanalysis.* New York: Viking, 1999.

Sears, William, and Martha Sears. *The Baby Book.* 1992. Reprint, New York: Little, Brown, 2003.

Seay, B., B. K. Alexander, and H. F. Harlow. "Maternal Behavior of Socially Deprived Rhesus Monkeys." *Journal of Abnormal and Social Psychology* 69 (1964): 345–54.

Seay, B., E. Hansen, and H. F. Harlow. "Mother-Infant Separation in Monkeys." *Journal of Child Psychology and Psychiatry and Allied Disciplines* 3 (1962): 123–32.

Senn, Milton J. E., ed. *Symposium on the Healthy Personality.* New York: Macy Foundation, 1950.

Sewell, William H. "Infant Training and the Personality of the Child." *American Journal of Sociology* 58 (1952): 150–59.

Shields, Stephanie A. *Speaking from the Heart: Gender and the Social Meaning of Emotion.* Cambridge: Cambridge University Press, 2002.

———. "To Pet, Coddle, and 'Do For': Caretaking and the Concept of Maternal Instinct." In *In the Shadow of the Past: Psychology Portrays the Sexes*, edited by Miriam Lewin, 256–73. New York: Columbia University Press, 1984.

Sidowski, J., and D. B. Lindsley. "Harry Frederick Harlow." *National Academy of Sciences Biographical Memoirs* 58 (1989): 219–57.

Smith, Lauren H. "Edward A. Strecker, M.D.: A Biographical Sketch." *American Journal of Psychiatry* 101 (July 1944): 9–11.

Sober, Elliott. "Evolution, Population Thinking, and Essentialism." *Philosophy of Science* 47 (1980): 350–83.

Solovey, Mark. "Project Camelot and the 1960s Epistemological Revolution: Rethinking the Politics-Patronage-Social Science Nexus." *Social Studies of Science* 31 (2001): 171–206.

———. *Shaky Foundations: The Politics-Patronage-Social-Science Nexus in Cold War America.* New Brunswick, NJ: Rutgers University Press, 2013.

Solovey, Mark, and Hamilton Cravens, eds. *Cold War Social Science: Knowledge Production, Liberal Democracy, and Human Nature.* New York: Palgrave Macmillan, 2012.

Sorokin, Pitirim A., and Robert C. Hanson. "The Power of Creative Love." In *The Meaning of Love*, edited by Ashley Montagu. New York: Julian Press, 1953.

Spitz, René A. "Hospitalism: An Inquiry into the Genesis of Psychiatric Conditions in Early Childhood." *Psychoanalytic Study of the Child* 1 (1945): 53–74.

———. "Anaclitic Depression: An Inquiry into the Genesis of Psychiatric Conditions in Early Childhood, II." *Psychoanalytic Study of the Child* 2 (1946): 313–42.

———. *Grief: A Peril in Infancy.* 1947. Film available at the Archives of the History of American Psychology and the US National Library of Medicine, Bethesda, MD.

———. "The Importance of the Mother-Child Relationship during the First Year of Life." *Mental Health Today* 7 (1948): 7–13.

———. "The Role of Ecological Factors in Emotional Development in Infancy." *Child Development* 20 (1949): 145–55.

———. *A Genetic Field Theory of Ego Formation: Its Implications for Pathology.* New York: International Universities Press, 1959.

———. "Discussion of Dr. Bowlby's Paper." *Psychoanalytic Study of the Child* 15 (1960): 85–94.

Spitz, René A., in collaboration with W. Godfrey Cobliner. *The First Year of Life: A Psychoanalytic Study of Normal and Deviant Development of Object Relations.* New York: International Universities Press, 1965.

Spitz, René A., and Katherine M. Wolf. "Hospitalism: A Follow-up Report." *Psychoanalytic Study of the Child* 2 (1946): 113–17.

Spock, Benjamin, and Michael B. Rothenberg. *Dr. Spock's Baby and Child Care.* 1945. Reprint New York: E. P. Dutton, 1985.

Sprengnether, Madelon. *The Spectral Mother: Freud, Feminism, and Psychoanalysis.* Ithaca, NY: Cornell University Press, 1990.

Stearns, Peter N. *Anxious Parents: A History of Modern Childrearing in America.* New York: New York University Press, 2003.

Stearns, Peter N., and Jan Lewis, eds. *An Emotional History of the United States.* New York: New York University Press, 1998.

Stearns, Peter N., and Carol Z. Stearns. "Emotionology: Clarifying the History of Emotions and Emotional Standards." *American Historical Review* 90 (1985): 813–36.

Sternglanz, Sarah Hall, and Alison Nash. "Ethological Contributions to the Study of Human Motherhood." In *Different Faces of Motherhood,* edited by Beverley Birns and Dale F. Hay, 15–46. New York: Plenum, 1988.

Stone, L. Joseph. "A Critique of Studies of Infant Isolation." *Child Development* 25 (1954): 9–20.

Strecker, Edward A. "Motherhood and Momism—Effect on the Nation." *University of Western Ontario Medical Journal* 16 (1946): 59–77.

———. "Presidential Address." *American Journal of Psychiatry* 101 (1944): 1–8.

———. *Their Mothers' Sons: The Psychiatrist Examines an American Problem.* Philadelphia: Lippincott, 1946.

———. "What's Wrong with American Mothers?" *Saturday Evening Post,* October 26, 1946, 14–15, 85–99, 102, 104.

Sulloway, Frank. *Freud: Biologist of the Mind.* Cambridge, MA: Harvard University Press, 1992.

Sulman, Michael. "The Humanization of the American Child: Benjamin Spock as a Popularizer of Psychoanalytic Thought." *Journal of the History of the Behavioral Sciences* 9 (1973): 258–65.

Suomi, Stephen J., and Harry F. Harlow. "Abnormal Social Behavior in Young Monkeys." In *Exceptional Infant: Studies in Abnormalities*, vol. 2, edited by Jerome Hellmuth, 483–529. New York: Brunner Mazel, 1971.

Suomi, S., and H. A. LeRoy. "In Memoriam: Harry F. Harlow (1905–81)." *American Journal of Primatology* 2 (1982): 319–42.

Tanner, J. M., and Barbel Inhelder, eds. *Discussions on Child Development*. 4 vols. in 1. London: Tavistock, 1971.

Taschwer, Klaus. "Von Gänsen und Menschen: Über die Geschichte der Ethologie in Österreich und über ihren Protagonisten, den Forscher, Popularisator und Ökopolitiker Konrad Lorenz." In *Wissenschaft, Politik und Öffentlichkeit, von der Wiener Moderne bis zur Gegenwart*, edited by Mitchell G. Ash and Christian H. Stifter, 331–51. Vienna: WUV, 2002.

Taschwer, Klaus, and Benedikt Föger. *Konrad Lorenz: Biographie*. Vienna: Paul Zsolnay, 2003.

Terry, Jennifer. "'Momism' and the Making of Treasonous Homosexuals." In *"Bad" Mothers: The Politics of Blame in Twentieth-Century America*, edited by Molly Ladd-Taylor and Lauri Umansky, 169–90. New York: New York University Press, 1998.

Thorpe, W. H. *Learning and Instinct in Animals*. Cambridge, MA: Harvard University Press, 1956.

Tinbergen, Niko. "An Objectivist Study of the Innate Behaviour of Animals." *Bibliotheca Biotheoretica* 1 (1942): 39–98.

———. "Psychology and Ethology as Supplementary Parts of a Science of Behavior." In *Group Processes: Transactions of the First Conference, September 26–30, 1954, Ithaca, New York*, edited by Bertram Schaffner, 75–167. New York: Josiah Macy Jr. Foundation, 1955.

———. *The Study of Instinct*. Oxford: Clarendon Press, 1951.

———. "On Aims and Methods of Ethology." *Zeitschrift für Tierpsychologie* 20 (1963): 410–33.

———. "Watching and Wondering." In *Studying Animal Behavior: Autobiographies of the Founders*, edited by Donald A. Dewsbury, 431–63. 1985. Reprint Chicago: University of Chicago Press, 1989.

Tobach, Ethel, and Lester R. Aronson. "T. C. Schneirla: A Biographical Note." In *Development and Evolution of Behavior*, edited by Lester R. Aronson, Ethel Tobach, Daniel S. Lehrman, and Jay S. Rosenblatt, xi–xviii. San Francisco: Freeman, 1970.

Trotter, Robert J. "Human Behavior: Do Animals Have the Answer?" *Science News* 105 (1974): 274–79.

Uexküll, Jacob von. "A Stroll through the Worlds of Animals and Men: A Picture Book of Invisible Worlds." In *Instinctive Behavior: The Development of a*

Modern Concept, translated and edited by Claire H. Schiller, 5–80. New York: International Universities Press, 1957.

Valenstein, E. S., W. Riss, and W. C. Young. "Experiential and Genetic Factors in the Organization of Sexual Behavior in Male Guinea Pigs." *Journal of Comparative and Physiological Psychology* 48 (1955): 397–403.

Van der Horst, Frank C. P. *John Bowlby: From Psychoanalysis to Ethology*. Chichester, UK: Wiley-Blackwell, 2011.

Van der Horst, Frank C. P., Helen LeRoy, and Renée Van der Veer. "'When Strangers Meet': John Bowlby and Harry Harlow on Attachment Behavior." *Integrative Psychological and Behavioral Science* 42 (2008): 370–88.

Van der Horst, Frank C. P., and René Van der Veer. "Separation and Divergence: The Untold Story of James Robertson's and John Bowlby's Theoretical Dispute on Mother-Child Separation." *Journal of the History of the Behavioral Sciences* 45, no. 3 (2009): 236–52.

Van Dijken, Suzan. *John Bowlby: His Early Life; a Biographical Journey into the Roots of Attachment Theory*. London: Free Association Books, 1998.

Vicedo, Marga. "Mother Love and Human Nature: A History of the Maternal Instinct." PhD diss., Harvard University, 2005.

———. "The Father of Ethology and the Foster Mother of Ducks: Konrad Lorenz as Expert on Motherhood," *ISIS* 100 (2009): 263–291.

———. "Mothers, Machines, and Morals: Harry Harlow's Work on Primate Love from Lab to Legend." *Journal of the History of the Behavioral Sciences* 45, no. 3 (2009): 193–218.

———. "The Evolution of Harry Harlow: From the Nature to the Nurture of Love." *History of Psychiatry* 21, no. 2 (2010): 1–16.

———. "Cold War Emotions: Mother Love and the War over Human Nature." In *Cold War Social Science: Knowledge Production, Liberal Democracy, and Human Nature*, edited by Mark Solovey and Hamilton Cravens, 233–49. New York: Palgrave Macmillan, 2012.

———. "The Objective Eye and the Subjective I in the Study of Animal Behavior." *Studies in the History and Philosophy of Science*. Forthcoming.

———. "Outside or Inside the Animal? Konrad Lorenz on Intuition and Empathy in the Study of Animal Behavior." *Studies in the History and Philosophy of Biology*. Forthcoming.

———. "Lehrman versus Lorenz: Facts, Semantics, and Politics in the Study of Animal Behavior." Unpublished manuscript.

———. "The Making of Attachment: The Theoretician, the Experimenter, and the Social Worker." Unpublished manuscript.

Watson, John B. "Instinctive Activity in Animals." *Harper's Magazine* 24 (1912): 376–82.

———. *Behavior: An Introduction to Comparative Psychology*. New York: Henry Holt, 1914.

————. *Psychology from the Standpoint of a Behaviorist*. Philadelphia: Lippincott, 1919.

Watson, John B., and J. J. B. Morgan. "Emotional Reactions and Psychological Experimentation." *American Journal of Psychology* 28 (1917): 163–74.

Weidemann, Doris. *Leben und Werk von Therese Benedek, 1892–1977: Weibliche Sexualität und Psychologie des Weiblichen*. Frankfurt: Peter Lang, 1988.

Weidman, Nadine. "Popularizing the Ancestry of Man: Robert Ardrey and the Killer Instinct," *ISIS 102* (2011): 269–299.

————."An Anthropologist on TV: Ashley Montagu and the Biological Basis of Human Nature, 1945–1960." In *Cold War Social Science: Knowledge Production, Liberal Democracy, and Human Nature*, edited by Mark Solovey and Hamilton Cravens, 215–32. New York: Palgrave Macmillan, 2012.

Weiss, Jessica. *To Have and to Hold: Marriage, the Baby Boom and Social Change*. Chicago: University of Chicago Press, 2000.

Wilson, Sloan. *The Man in the Gray Flannel Suit*. New York: Simon and Schuster, 1955.

Winston, Andrew S. "'As His Name Indicates': R. S. Woodworth's Letters of Reference and Employment for Jewish Psychologists in the 1930s." *Journal of the History of the Behavioral Sciences* 32, no. 1 (1996): 30–43.

————. "Mythologized History of Psychology as Social Practice: The Uses of 'Environmentalist Hegemony' Narratives." In *History of Psychology and Social Practice,* edited by A. C. Brock and J. Louw, 83–95. Special issue of *Social Practice/Psychological Theorizing*, 2007.

Witmer, Helen Leland. "Introduction." In *Symposium on the Healthy Personality*, edited by Milton J. E. Senn, 13–14. New York: Macy Foundation, 1950.

Witmer, Helen Leland, and Ruth Kotinsky, eds. *Personality in the Making: The Fact-Finding Report of the Midcentury White House Conference on Children and Youth*. New York: Harper and Row, 1952.

Wittenberg, Stephen M., and Lewis M. Cohen. "Max Schur, M.D., 1897–1969." *American Journal of Psychiatry* 159 (2002): 216.

Wittenborn, J. Richard, assisted by Barbara Myers. *The Placement of Adoptive Children*. Springfield, IL: Charles C. Thomas, 1957.

Wootton, Barbara. "A Social Scientist's Approach to Maternal Deprivation." In World Health Organization, *Deprivation of Maternal Care: A Reassessment of Its Effects,* 63–74. Geneva: World Health Organization, 1962.

World Health Organization. *Deprivation of Maternal Care: A Reassessment of Its Effects*. Geneva: World Health Organization, 1962.

Wortis, Rochelle Paul. "The Acceptance of the Concept of the Maternal Role by Behavioral Scientists: Its Effects on Women." *American Journal of Orthopsychiatry* 41, no. 5 (1971): 733–46.

Wylie, Philip. *Generation of Vipers*. New York: Rinehart, 1942.

————. *An Essay on Morals*. New York: Rinehart, 1947.

————. *The Magic Animal*. New York: Doubleday, 1968.

Young-Bruehl, Elisabeth. *Anna Freud: A Biography,* 2nd ed. New Haven, CT: Yale University Press, 2008.

Zegans, Leonard S. "An Appraisal of Ethological Contributions to Psychiatric Theory and Research." *American Journal of Psychiatry* 124 (1967): 729–39.

Zuckerman, Michael. "Dr. Spock: The Confidence Man." In *The Family in History*, edited by Charles Rosenberg, 179–207. Philadelphia: University of Pennsylvania Press, 1975.

Archives Consulted

Mary Ainsworth Papers, Archives of the History of American Psychology (hereafter AHAP), Center for the History of Psychology, University of Akron, Akron, OH.

Frank Beach Papers, AHAP, Akron, OH.

John Bowlby Papers, Western Manuscripts and Archives, Wellcome Library, London.

Leonard Carmichael Papers, Manuscripts Department, American Philosophical Society Library, Philadelphia.

Erik Erikson Papers, Houghton Library, Harvard University, Cambridge, MA.

Anna Freud Papers, Sigmund Freud Collection, Manuscript Division, Library of Congress, Washington, DC.

Harry Harlow Papers, Harlow Primate Laboratory, University of Wisconsin-Madison, Madison, WI.

Donald O. Hebb Papers, McGill University Archives, Montreal.

Robert Hinde Papers, Personal Collection, St. John's College, Cambridge University, Cambridge.

Daniel Lehrman Papers, Institute of Animal Behavior, Rutgers University, Newark, NJ.

David Levy Papers, Oskar Diethelm Library, DeWitt Wallace Institute for the History of Psychiatry, Weill Cornell Medical College, New York City.

Konrad Z. Lorenz Papers, Konrad Lorenz Institute for Evolution and Cognition Research, Altenberg, Austria.

T. C. Schneirla Papers, AHAP, Akron, OH.

René Spitz Papers, AHAP, Akron, OH.

William Homan Thorpe Papers, Cambridge University Library, Department of Manuscripts and University Archives, Cambridge.

Nikolaas Tinbergen Papers, Oxford University, Bodleian Library, Department of Special Collections and Western Manuscripts, Oxford.

William S. Verplanck Papers, AHAP, Akron, OH.

Philip Wylie Papers, Manuscripts Division, Department of Rare Books and Special Collections, Princeton University Library, Princeton, NJ.

Index

Page numbers in italics refer to images.